Contents

Acknowledgements		02
1	Introduction to Subject Specific Guidelines	06
1.1	Purpose of the SSGs	06
1.2	The SSGs in context	06
1.3	SSGs and the issue of Complexity versus Maturity	06
1.4	Further reading	06
2	Introduction to the Asset Information SSG	08
2.1	How to use this SSG	08
2.2	What is Asset Information?	09
2.3	Asset Information Systems	10
2.4	Value of Asset Information	10
2.5	Scope and Boundaries	12
2.6	Target Sectors	12
2.7	Target audience	12
2.8	Definitions	13
2.9	Data Quality	13
2.10	Key Messages	13
3	Good Practice Approach for Asset Information Management	14
3.1	Purpose and Structure	14
3.2	Good practice approaches	14
3.3	Key Messages	17
4	The Asset Information Management System	18
4.1	Assess and manage risks	18
4.2	Collect or acquire information	18
4.3	Receive, use, transmit and transform information	18
4.4	Maintain information	18
4.5	Comply with legal and other requirements	20
4.6	Manage change	20
4.7	Archival, retention and disposal of information	20
4.8	Periodic Review and improvement of the information management system	20
4.9	Key Messages	20
5	Governance	22
5.1	Governance Processes	22
5.2	Meta-information	25
5.3	Responsibilities	25
5.4	Key Messages	25
6	Asset information strategy	28
6.1	What is an Asset Information Strategy?	28
6.2	Why does an organisation need an Asset Information Strategy?	28
6.3	Developing and Agreeing an Asset Information Strategy	29
6.4	Implementing the Asset Information Strategy	29
6.5	Updating the Asset Information Strategy	29
6.6	Key Messages	29
7	Standards, specifications and requirements	30
7.1	Why is Asset Information needed?	30
7.2	What Asset Information is needed?	32
7.3	Considerations during requirements gathering	32
7.4	The Asset Data Dictionary	32
7.5	Key Messages	33
8	Information lifecycle	34
8.1	Assess data	34
8.2	Improve data	38
8.3	Store data	45
8.4	Utilise data	46
8.5	Acquire new data	50
8.6	Archive data	51
8.7	Delete data	51
8.8	Sustain data quality	52
8.9	Key Messages	52
9	Monitoring	54
9.1	Data quality measures	54
9.2	Targets	54
9.3	Compliance	55
9.4	Presentation of Measures	55
9.5	Key Messages	55
10	Audit and Assurance	56
10.1	Audit	56
10.2	Assurance	57
10.3	Key Messages	57
11	Benchmarking	58
11.1	Why benchmark?	58
11.2	Benchmarking and change	58
11.3	What is 'Good Practice'?	58
11.4	Assessment guidelines and maturity modelling	58
11.5	Alignment with the IAM Self-Assessment Methodology Plus (SAM+)	60
11.6	Asset Information Subjects for assessment	62
11.7	Key Messages	62
12	The organisation and the people dimension	64
12.1	People	64
12.2	Key Messages	67
13	Process	68
13.1	Key Messages	68
14	Software	70
14.1	Getting the best from information technology	70
14.2	Enterprise Architecture	71
14.3	Information Architecture and the Data Dictionary	71
14.4	Data Mastering Strategy	72
14.5	The role of information technology in projects	74
14.6	Key Messages	76
15	Managing change	78
15.1	Introduction	78

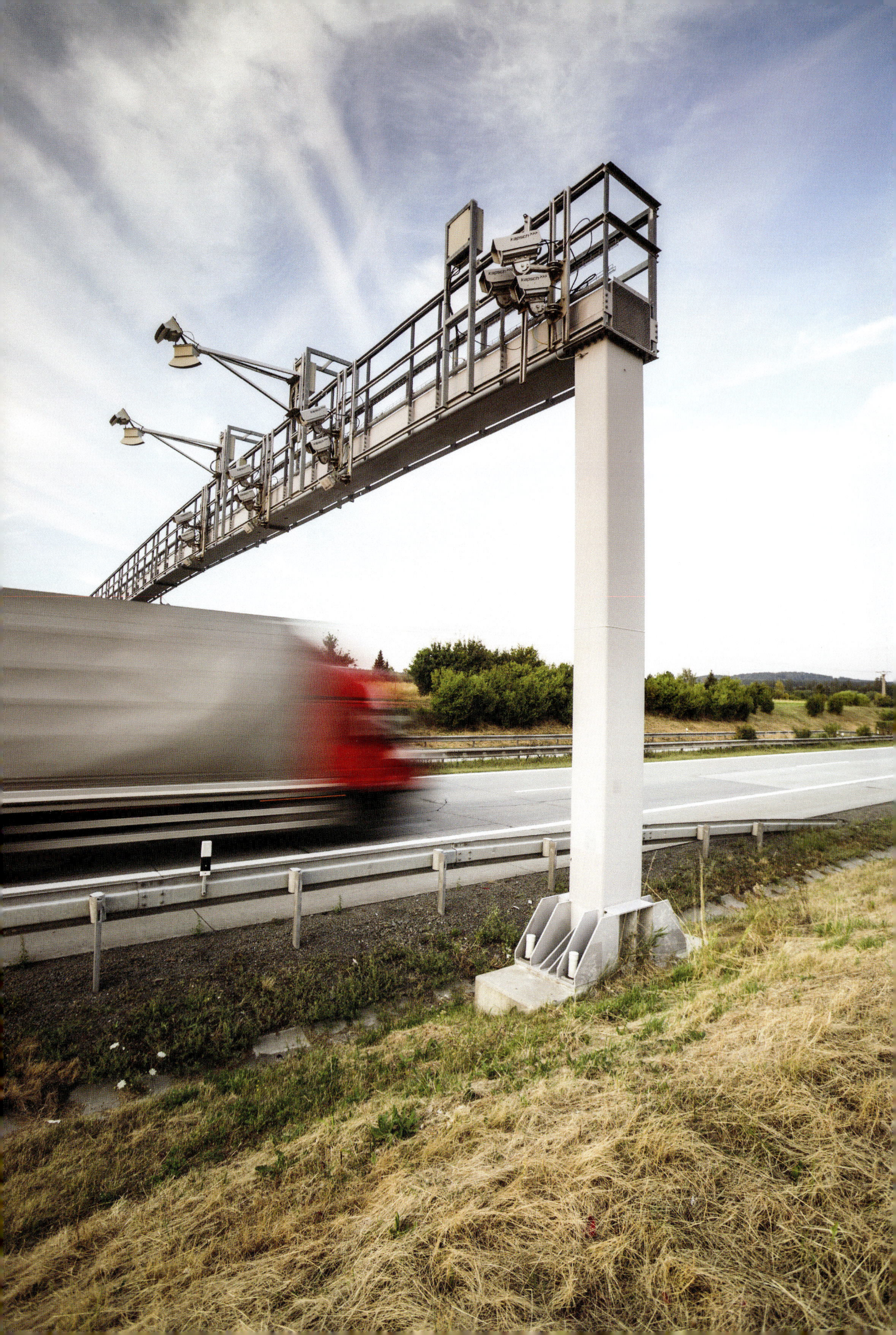

Contents

15.2	What is to change	78
15.3	Technology Considerations	79
15.4	Assessing Options and creating a Business Case	79
15.5	Preparing for the Change Programme	80
15.6	Programme Management	81
15.7	Creating the new Processes and Systems	81
15.8	Commissioning and changes to Processes and Systems	81
15.9	Culture Change and Change Management	82
15.10	Key Messages	82
16	Glossary of terms	84
17	Further reading	86
17.1	Bibliography	86
17.2	Publications	86
17.3	Useful web sites	86

About the IAM

This document is for guidance and information only. The Institute of Asset Management takes no responsibility for the usage or applicability of the Asset Information Guidelines for any selection, development, or implementation of new working processes, systems, or technologies, or alterations to how businesses use asset information to make decisions.

Our Objectives

- Advance for the public benefit the science and practice of Asset Management
- Promote and recognise high standards of practice and professional competence
- Generate widespread awareness and understanding of the discipline

Please contact us

Successful Asset Management requires a combination of skills, techniques and knowledge, particularly finance and we welcome engagement and collaboration with other expert bodies and interested individuals. Please visit us at www.theIAM.org

1 Introduction to Subject Specific Guidelines

This Subject Specific Guidance (SSG) is part of a suite of documents designed to expand and enrich the description of the Asset Management discipline as summarised in the IAM's document 'Asset Management – an Anatomy' (referred to throughout this document as *'The Anatomy'*).

The SSGs cover the 39 Subjects in The Anatomy directly as a 'one to one' (where a subject is very broad), or grouped (where subjects are very closely related).

1.1 Purpose of the SSGs

This document provides guidance for good asset management. It is part of a suite of Subject Specific Guidance documents that explains the 39 subject areas identified in "Asset Management – an Anatomy", also published by the Institute of Asset Management. These subject areas are also acknowledged by the Global Forum for Maintenance and Asset Management as the "Asset Management Landscape".

PAS 55 and ISO 55001 set out requirements which describe **what** is be done to be competent in asset management, however they don't offer advice on **how** it should be done. The SSGs are intended to develop the next level of detail for each subject in The Anatomy. They should therefore be read as **guidance**; they are not prescriptive, but rather intended to help organisations by providing a consolidated view of good practice, drawn from experienced practitioners across many sectors.

The SSGs include simple as well as complex solutions, together with real examples from different industries to support the explanatory text because it is understood that industries and organisations differ in scale and sophistication. In addition, they are at different stages of asset management; some may be relatively mature while others are at the beginning of the journey.

Accordingly, there is flexibility for each organisation to adopt their own 'fit for purpose' alternative practical approaches and solutions that are economic, viable, understandable and usable. The underlying requirement for continual improvement should drive progress.

1.2 The SSGs in context

The SSGs are a core element within the IAM Body of Knowledge and they have been peer reviewed and assessed by the IAM Expert Panel. They align fully with the IAM's values and beliefs that relate to both the development of excellence in the asset management discipline and provision of support to those who seek to achieve that level of excellence.

1.3 SSGs and the issue of Complexity versus Maturity

It is important to understand and contrast these terms. Put simply:
- The complexity of the business will drive the complexity of the solution required; and
- The maturity of the organisation will determine its ability to recognise and implement an appropriate solution.

A very mature organisation may choose a simple solution where a naive organisation may think that a complex solution will solve all its problems. In truth, there is no universal best practice in Asset Management – only good practice that is appropriate for the operating context of any particular organisation. What is good practice for one organisation may not be good practice for another.

For example, an organisation that is responsible for managing 100 assets, all in the same location, could use a spread sheet-based solution for an Asset Register and work management system. This is arguably good practice for that organisation. However, for a utility business with thousands of distributed assets, this is unlikely to represent a good practice solution. When reading the SSGs, the reader should have a view of the complexity and maturity of the organization, and interpret the guidance that is offered in that context.

1.4 Further reading

The Anatomy provides a starting point for development and understanding of an Asset Management capability and the SSGs follow on to support that further. However, the opportunity doesn't end there; the IAM provides a range of expert and general opinion and knowledge which is easily accessed by members through the IAM website.

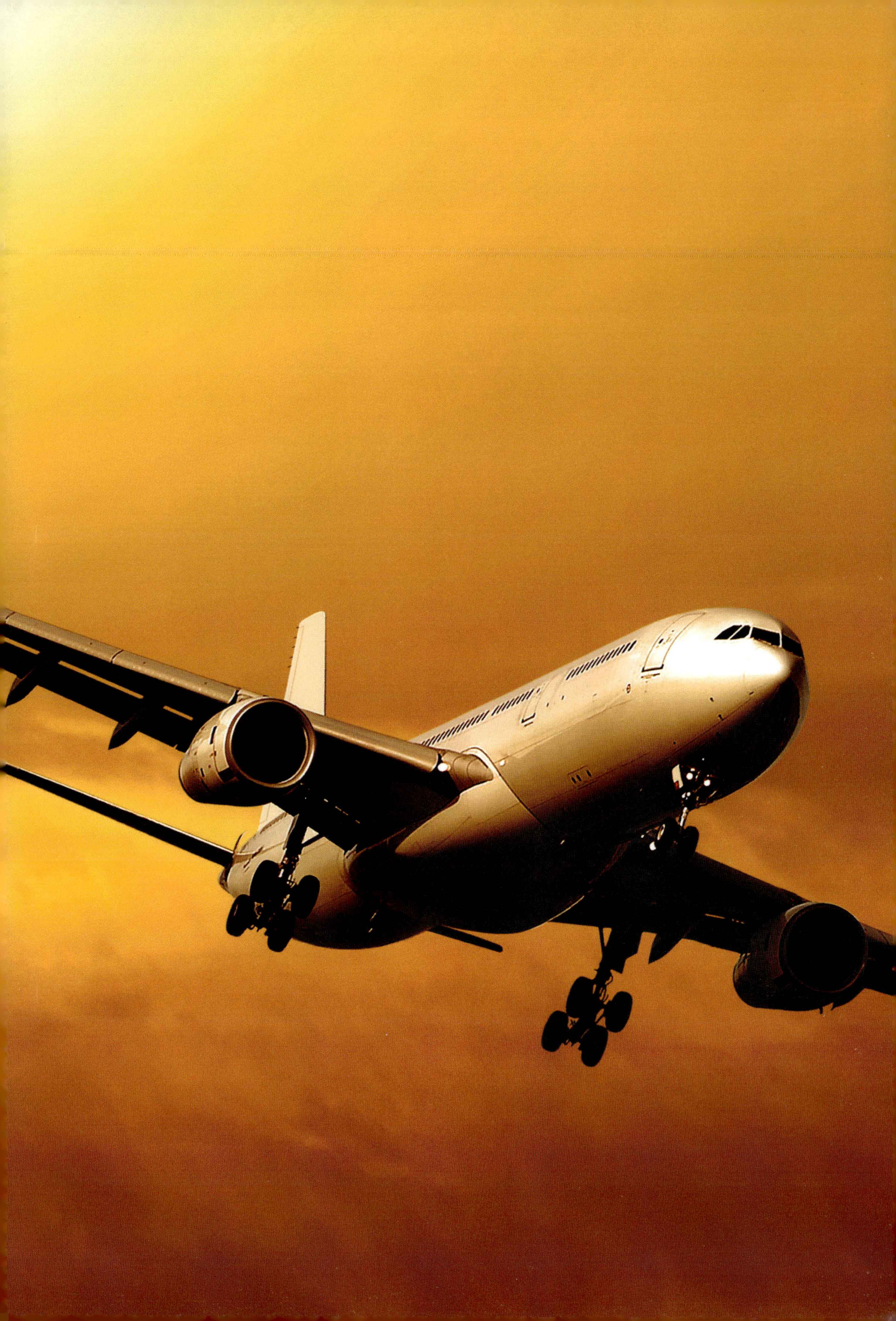

2 Introduction to the Asset Information SSG

This SSG has been developed in response to demand from industry for guidance in carrying out better management of asset information. It is applicable to any organisation where physical assets are a key or critical factor in achieving its business objectives and in achieving effective service delivery.

Many infrastructure-related businesses have invested heavily in asset information systems and data gathering to help manage their assets and improve their overall efficiency and performance. Such programmes typically include new software tools and modified working processes, and are aimed at providing better information to help the asset management function make better decisions. Despite very significant expenditure, many businesses complain that benefits have been slow to be delivered, are difficult to quantify, and their asset managers still say that they do not have access to the information they need.

The necessity for good asset information is growing rather than reducing. Requirements are becoming more sophisticated, and the number of stakeholders, and hence the complexity of collating and sharing information, is increasing. This trend is set to continue, given the increasingly stringent regulation of utilities, the increase in private finance initiatives to fund major capital programmes, and a greater collective understanding of asset management risk.

In spring 2009 the Institute of Asset Management (IAM) published a document entitled "Asset Information Guidelines". Subsequently, in August 2010 a draft Asset Information Quality Handbook was released detailing the many activities that contribute to the management of asset information quality, which was also well received by Members of the IAM.

In 2013 a review of both the Asset Information Guidelines and the Asset Information Quality Handbook lead to the development of this SSG as part of the IAM Subject Specific Guidelines (SSGs).

This SSG's intended audience is the community of people responsible for the management and utilisation of information associated with assets. This group typically includes Asset Management, Engineering, Operations, Finance, and IT departments. However, it also recognises that the understanding of asset information management in an asset management context is an emerging discipline and it is therefore written to allow 'non-practitioners' to better understand the key features of good practice asset information management.

It is emphasised that this document is not intended to provide guidance on Information Technology or Software Engineering.

There are other good practice guides available for the procurement, design, development and implementation of software. However, key areas of overlap and interface are covered for clarity. Importantly, the pretext for this document is the understanding that the most pressing improvements required for asset information concern better process and utilisation, and are not necessarily dependent on technology or software.

There are many publications detailing generic approaches to managing data and information quality. This SSG adopts many of these techniques and concepts and applies them to the field of asset management.

2.1 How to use this SSG

This SSG has been developed as a key reference source. Due to the breadth of topics covered and the fact that each organisation will be starting from a different point, it is suggested that Section 3 is used a starting point and that this should link to the Section relevant to a reader's interest.

Each Section is intended to be reasonably self-contained, so it should be possible to gain value by reading the Sections relevant to a reader's interest.

As understanding develops, it is likely that re-reading Sections may be useful to reinforce understanding.

Good practice approaches to asset management, as detailed in PAS55 or ISO 55000, require the establishment of an asset management system to direct, coordinate and control asset management activities. For clarity, the term 'system' refers to the business system and not software systems used to undertake asset management.

Asset information provides support both to individual parts of the asset management system and enables better coordination between parts. As such, good management of asset information is a key enabler for good asset management and should also be managed as a system.

2 Introduction to the Asset Management SSG

2.2 What is Asset Information?

Asset information is a combination of data about physical assets used to inform decisions about how they are managed both for short term operational purposes and for long term strategic planning. The term covers:

- **Asset register/inventory** – For most organisations the Asset Register (or Asset Inventory) is central to the effective management of assets. An Asset Register can be a single data store or a number of linked and related data stores, depending on the organisation and the nature of their assets. Typically, the Asset Register will be linked to a number of other data sets that together support the effective management of assets and other business objectives. Asset attributes in the Asset Register tend to remain static over time
- **Asset history** – key information surrounding the creation, operation, maintenance and retirement of an asset along with the activities undertaken. This information tends to change over time, so time/date recording of history is important
- **Documents** – Both generic suppliers documents and system/asset specific text based documents
- **Design information** – 2D and 3D CAD data, design calculations
- **Images and multimedia**

Some of this information may be subjective, for example, assessments of asset condition grades

> **Note:** There is not a rigid boundary around asset information; therefore it should not be managed in isolation. Asset information may be used for other business purposes, conversely, data from other business purposes may be used in asset management.

Good asset information enables better decisions to be made, such as determining the optimal maintenance or renewal frequency for an asset. That decision may be based on information regarding the asset's location, condition, probability and consequence of failure, work option specifications and costs, constraints such as resource availability, and other business priorities, such as compliance with regulatory requirements.

The nature of such information varies and overlaps considerably. Understanding what information is required within an asset management context, how it should be collected, stored, and analysed, and how this process should be repeated or amended over time, is a complex question for all asset intensive businesses.

Whilst technology undoubtedly plays an important role, asset information management professionals consider the challenge in a broader context, such as identifying:

- What questions do we need to answer to effectively manage our assets? What asset information do we need to answers these questions and why?
- How do we measure and assure the quality of our information, and what risk do we carry if the quality is poor or the content incorrect?
- How should we control the creation, management and maintenance of asset information?
- How do we ensure that our people and stakeholders understand the importance of asset information and their role in its lifecycle?
- How to ensure that all relevant stakeholders and staff treat data and the outputs of data analysis correctly?
- How to ensure that staff provide, use, store and manage data appropriately to their role?
- Is there an audit trail to justify approaches to decision making?
- What use should be made of technology and software, and do we need to change what we already have?
- How do we get better value from our existing people, software, management systems and processes?

The Institute of Asset Management's BSI PAS 55 publication[3] was a significant step forward in addressing some of these issues, as it identifies and clarifies the requirements for asset information and systems within an asset management context. These guidelines seek to supplement this specification by providing guidance on how to meet those requirements.

The Bibliography also references a number of publications[4,5,6] with a variety of approaches that the reader may find helpful that are not re-iterated here.

2.3 Asset Information Systems

Organisations should establish and maintain systems that manage asset information. The systems should be designed to provide sufficient support and information to meet the organisation's asset management decision making objectives.

The scope of asset information is significant and is directly related to the scope of the asset management regime. The conceptual diagram illustrating the asset management regime is shown in Figure 1. The item 'Asset Knowledge Enablers' refers to asset information.

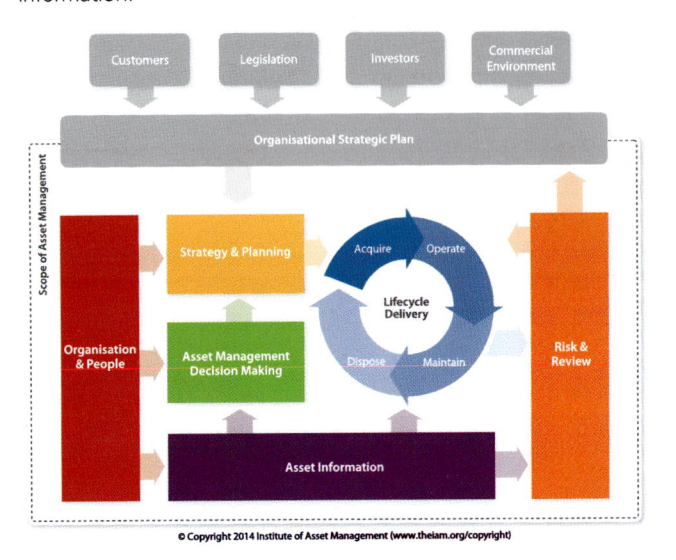

Figure 1 Scope of Asset Management

While the key components, considerations and definitions of asset information management are generic across industry sectors, the specific requirements of every business will vary and will depend on its size, nature and asset management strategy. Each business should choose and adapt information processes for their own requirements. There is an extensive range of Asset Management Information Systems available. These systems may be comparatively simply based around a register of assets but can be very comprehensive, including works scheduling systems helping with operational maintenance and advanced decision support tools that model the deterioration of asset systems over the system lifecycle. Some of the more advanced asset information systems are capable of predicting system failures by simulating the behaviours of assets under stress. Additionally, locally developed systems may be a key part of the overall approach.

The information needs, information systems design and software architecture is developed by the business and it is up to the business to define how their information system works. Unless the content of the asset information system is managed appropriately then the businesses decision making capability will be impaired.

For example:

- When maintaining assets – too much or too little maintenance may be carried out resulting in unnecessary work or costly asset failures. Maintenance records need to be comprehensive and accurate so Engineers can understand the performance of the asset during its life;
- When setting investment requirements/ Capital expenditure planning – insufficient information is available on asset condition and system performance. Investment decisions may only be made on lapsed time as opposed to actual asset condition;
- When responding to alarms and operational incidents – too much time and effort is carried out with reactive work, when asset defects and failures are not reported properly;
- When managing logistics – incorrect parts have been sent to repair the asset causing unnecessary delays;
- When upgrading – an incomplete Health and Safety File arising from a capital works project not returning the information needed to carry out an asset upgrade. This causes delay, involving unnecessary site surveys;
- When under review – not being able to meet regulatory targets and therefore not being able to take advantage of incentive mechanisms.

Having an effective asset information management system is a key component of asset management. Such a system ensures that the right information is available to the right users at the right time to support business objectives.

2.4 Value of Asset Information

2.4.1 Cost Efficiencies

Recent studies[7,8] have shown that asset information has a very significant effect on the efficiency and performance of asset intensive businesses. Organisations operating efficient asset information processes have been found to indirectly spend around 20% of their total annual budget (OPEX and CAPEX) on asset information. Businesses with poor asset information processes were found to spend even more, typically as much as 25%.

These figures may appear high, but include the cost of preparing and recording asset information when during maintenance and renewal, and significant hidden costs, such as management and staff time spent searching for information, collating and processing it from multiple sources and formats, and perhaps repeating or duplicating the process across multiple business departments. The majority of the cost is therefore embedded in the core business processes that are dependent on asset information, as illustrated in Figure 2 below. Improving the efficiency of how asset information is managed within the business therefore offers a significant opportunity for savings which has been estimated to be in the order of 1 – 5% of total business expenditure.

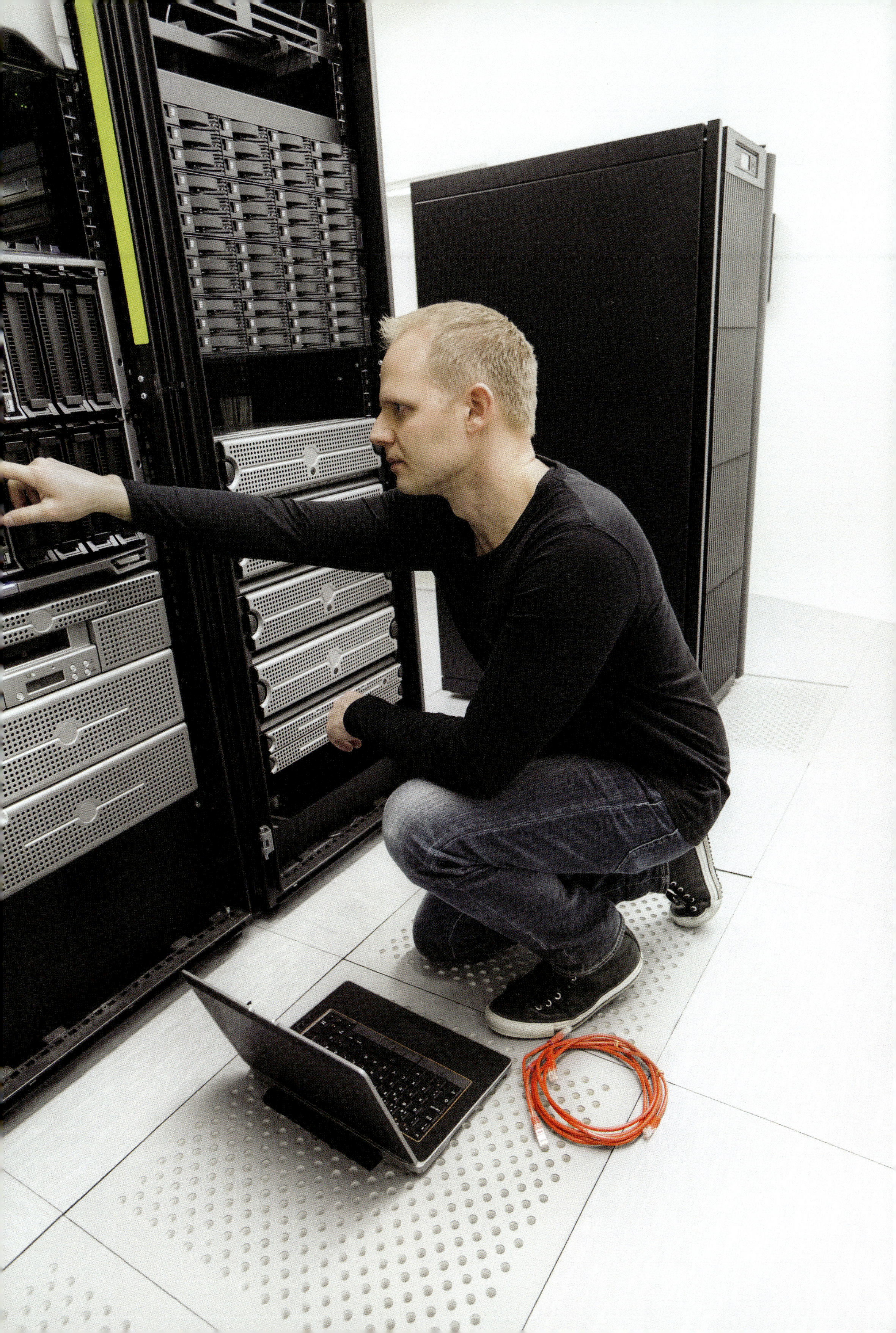

2.4.2 Expenditure effectiveness

An even greater benefit can be realised if asset information is used effectively to inform decision making on business expenditure profiles, such as capital programmes to improve asset serviceability, or best whole life cost decisions regarding maintenance and renewal choices. Put simply, the appropriate utilisation of asset information will enable the right work to be done in the right place at the right time. Conversely, the lack of reliable asset information in this context can result in poor or suboptimal decision making which can expose the business to unnecessary cost or risk and adversely affect overall business performance.

2.5 Scope and Boundaries

Effective asset information management requires several different parts of the organisation to work together, and the Asset Management and Information Technology/Systems functions in particular. Similarly, the governance and utilisation of asset information requires effective business processes to be established, and competent people to work within those processes. The rigour of the processes becomes even more important when service partner organisations provide significant parts of the delivery of core processes, and hence are involved in the collection of information about them. Processes and competencies are significant topics and disciplines in their own right, and have been covered by other good practice guidelines, such as BSI PAS 55 and the Institute of Asset Management's Competences Framework 9.

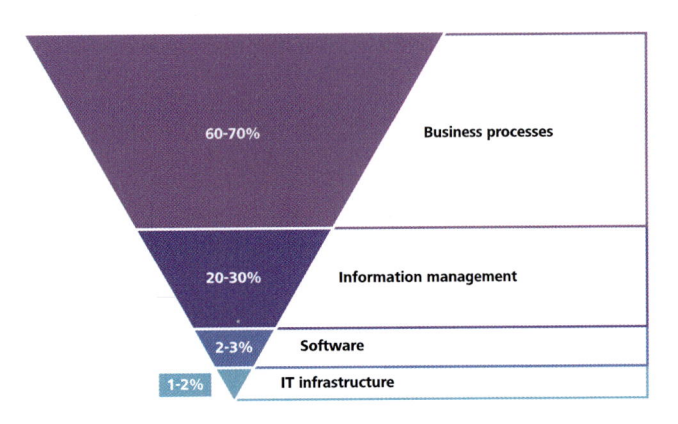

Figure 2 - Operational cost of Asset Information

This SSG is focused on asset information within an asset management context. The requirements for asset information as specified in BSI PAS 55 are used as a focus for information requirements. This document elaborates on those requirements and provides guidance on how they can be met, including illustrative case studies where appropriate. Information and technology are considered in two contexts. Firstly in recognising that such tools will only deliver asset management value if they are used to influence decision making, and hence their scope and cost justification should be considered relative to the value of those decisions. Secondly, in recognising that the emergence of new technology enables asset managers to gain information and asset knowledge that was previously not viable, but which could deliver significant asset management benefits, such as automated inspection systems or hand held field units linked to a central work management system. These opportunities should be considered in the context of innovation and continuous improvement.

Other corporate information systems, such as customer billing, payroll, or personnel software are business critical but typically contain information that is not primarily used for asset management. Although such data stores are excluded from this scope, the asset information they contain must be included and taken into account in developing asset information systems.

2.6 Target Sectors

This SSG is relevant to all major asset intensive sectors including, but not limited to:
- Water treatment, supply and sewage treatment
- Electricity generation, transmission and distribution
- Gas transmission and distribution
- Rail including mainline, metropolitan and light rail networks
- Airports, ports and freight handling facilities
- Highways and roads
- Petrochemical and process industries
- Property

The SSG recognises that organisations are likely to fall into three broad categories in which to consider how to deliver improved asset information, namely:
- For a new venture where there is opportunity to create new asset information and systems;
- Where there is opportunity to overhaul information or systems for an existing asset management enterprise, such as after an acquisition or major reorganisation; and
- Where there is opportunity to improve the use or quality of existing asset information or systems, which is applicable in any organisation.

Each of these circumstances is very different in terms of the freedom and flexibility to make asset information improvements. Key guidance points are made relative to these distinctions in each section of the document where possible.

2.7 Target audience

The target audience of this SSG are typical asset information users, for example:
- Business Area Sponsors – those that are responsible for promoting the development and use of information systems that manage asset information

2 Introduction to the Asset Management SSG

- Data and Process Owners – those that are responsible for making asset management decisions and running key business processes
- Data Stewards – those that are responsible for managing the production of asset information
- Information Governance and support – Information auditors and information assurance roles
- Data suppliers and consumers – providers and end users of the asset information

The responsibilities listed above do not necessarily align with organisational roles.

2.8 Definitions

Asset information is a combination of data about physical assets used to inform decisions about how they are managed. Numerous definitions of data and information exist. For the purposes of this document the definitions are used as follows:

- Data - numbers, words, symbols, pictures, etc. without context or meaning, i.e. data in a raw format, e.g. 25 metres;
- Information - a collection of data expressed with a supporting context e.g. The span of the bridge is 25 metres;
- Record – evidence in the form of information representing an account of something that has occurred e.g. a maintenance record detailing an item of work being carried out;
- Knowledge - a combination of experience, values, information in context, and insight that form a basis for decisions making. It refers to the process of comprehending, comparing, judging, remembering, and reasoning;
- Information Management - the means by which an organisation maximises the efficiency with which it plans, collects, organises, uses, controls, stores, disseminates, and disposes of its Information, and through which it ensures that the value of that information is identified and exploited to the maximum extent possible. The aim has often been described as getting the right information to the right person, in the right format and medium, at the right time; and
- Information Technology - the technology used (e.g. applications and software) to support business functions and processes.

These definitions, plus many other relevant terms are included in the Glossary of this SSG.

2.9 Data Quality

Several dimensions can be applied when measuring data quality. Each dimension can be associated with a different qualitative context for example:

- Accuracy – the record is correct in all details and is a true record of the entity it represents;
- Completeness – the record having all or the necessary attribute values relative to its intended purpose. Additionally, all entities of a particular class or type are recorded;
- Validity – data conforms to all standards expected;
- Consistency – an entity that is represented in more than one data store can easily be matched;
- Uniqueness – a single representation exists for each physical entity;
- Timeliness – data is easily accessed when required and is up to date.

How these metrics are interpreted and applied is dependent upon how the organisation manages its information systems. More details on data quality dimensions is provided in Section 8.1

2.10 Key Messages

i. The Asset Information SSG provides practical advice for asset management practitioners, but the document is not a prescriptive standard
ii. The SSG is complimentary to BSI PAS55/ISO 55000
iii. The SSG does not specify good practice for IT technology and software
iv. The specific requirements of each organisation will vary and will depend on its size, nature of assets and asset management strategy. They will also each be at a different level of maturity with different objectives, so will need to adapt approach detailed in this SSG
v. Asset information is a key enabler for effective asset management and should be viewed as an asset in its own right
vi. Information to support asset management activities will typically also support other organisational processes/ priorities and should not be viewed in isolation
vii. Asset information needs to be managed and not just created; managing asset information incurs costs, however, failure to manage information effectively will lead to far higher costs
viii. The requirements for asset information typically change over time so will need to be reassessed periodically
ix. One of the greatest risks to good asset information is that it easily becomes unreliable or degraded through failure to maintain it and the cost of recovering its reliability may be prohibitive
x. The quality of asset information needs to be actively managed and used to minimise costs, maximise benefits and prevent long degradation in quality
xi. Data quality is typically assessed using the metrics of accuracy, completeness, validity, consistency, uniqueness and accessibility

3 Good Practice Approach for Asset Information Management

Appropriate quality data and information are key to the management of any asset management business. It provides support for the asset management activity itself, for policy making and for supporting both operational and strategic business decisions.

3.1 Purpose and Structure

The following diagram illustrates the scope of the "asset information management system" – this is used as the core reference on which the structure of this document has been based. The word system is used to represent the management systems that are established and does not refer to software. The word 'governance' in this context, refers to the disciplines involved in managing and controlling data and information. Each chapter represents a "good practice" highlighted in this diagram and is further extended to include key messages.

Ultimately an organisation needs to consider what it deems as essential for defining its own good practice suitable to the scale and nature of its asset management operation.

3.2 Good practice approaches

The **Asset Information Management System** is established to define and manage the use of asset information by the organisation as a key component, and in support of, the Asset Management System. The Asset Information Management

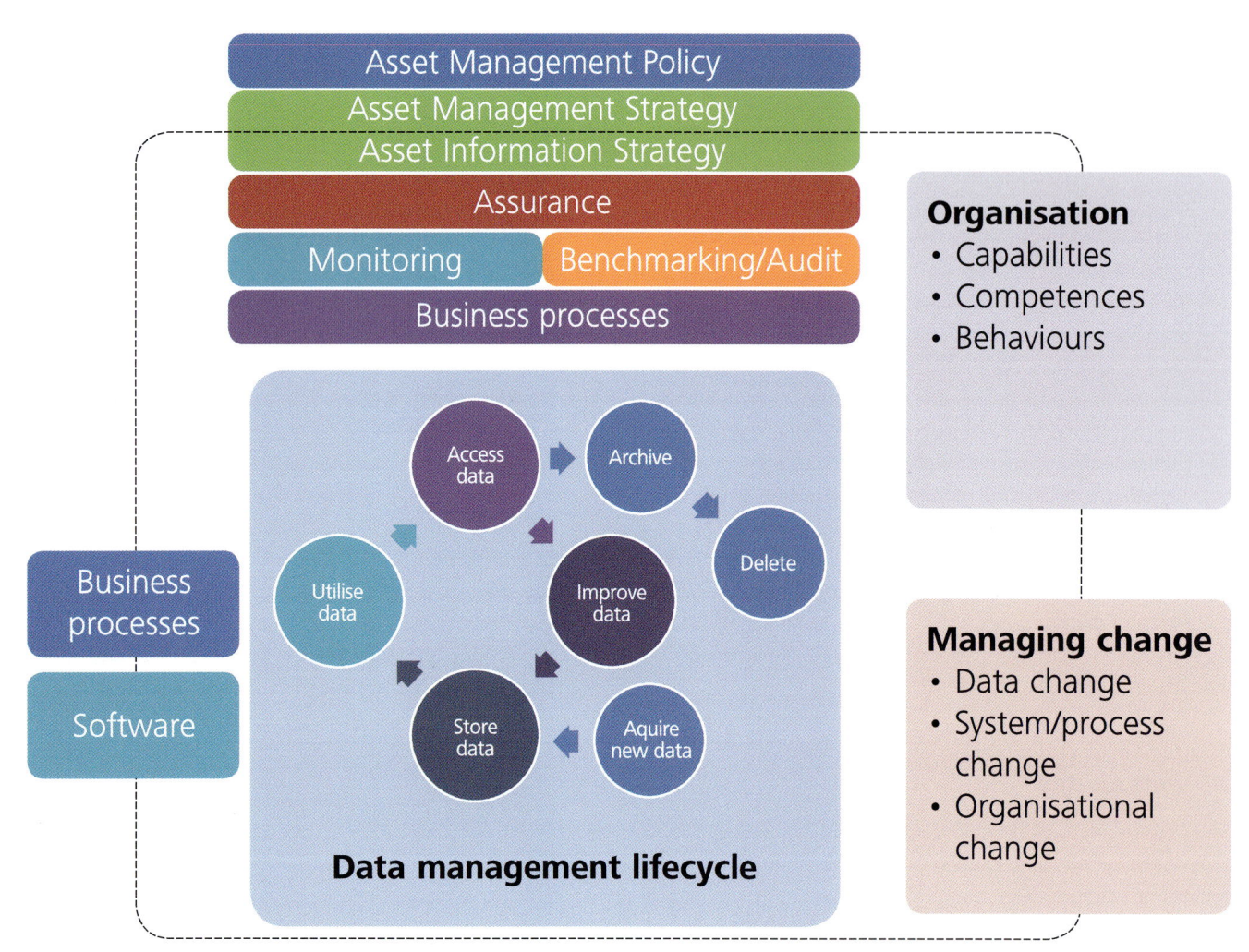

Figure 3 – The Asset Information Management System

3 Good Practice Approach for Asset Information Management

Section	Scope	Good Practice Examples
4	Asset Information Management System	• An overall system is established to control asset information • The system receives senior stakeholder involvement • The system is reviewed on a periodic basis to ensure alignment to asset management objectives
5	Governance	• A governance group of senior stakeholders meets regularly to oversee the usage and management of asset information • The governance group ensures that an Asset Information Strategy is agreed and that ongoing activities conform to the strategy
6	Asset Information Strategy	• The AIS is clearly aligned to the Asset Management Strategy • The current state of asset information management is stated • Agreed target end state is clearly stated • A roadmap for delivery of strategy is defined
7	Standards, Specifications and Requirements	• Clear specifications and requirements for asset information are agreed and defined • An Asset Data Dictionary stating these requirements is available to all relevant staff • Internal and external Standards that must be complied with are clearly stated
8	Information Lifecycle	• The current state of asset data has been assessed • Required improvement actions have been defined and agreed • Data storage is clearly defined and conforms to the organisations approach to master data • Effective processes exist to ensure that new asset data is acquired and stored efficiently • Clear rules have been defined for the archiving and deletion of data
9	Monitoring	• A comprehensive set of data quality rules will have been defined • The data quality rules include assessment of the accuracy, validity, completeness, uniqueness, consistency and timeliness of data • Regular monitoring of data stores against these data quality rules is undertaken
10	Audit and Assurance	• Audits are used to provide an in-depth assessment of the data, processes and standards in use. • A specific responsibility to provide ongoing assurance of asset information processes will have been established to provide assurance that company standards are being followed and that appropriate controls are in place to maintain compliance
11	Benchmarking	• Periodic assessments of the performance of the organisation against relevant external comparator organisations are undertaken • The results of benchmarking may be used to instigate changes to strategy, standards or requirements
12	Organisation	• The organisation has been structured to support good asset management and asset information practices • The capability of the organisation is clearly defined and understood • The individual competences required to support good asset management/ asset information management are defined and influence the training/development plans of individual staff members • Staff understand the importance of data and how their behaviour can affect data quality
13	Business Process	• Processes are clearly defined, mapped and followed • The data inputs and outputs from process activities are clearly understood • Critical data required to support business process is understood • Process metrics defined and utilised appropriately
14	Software	• A clear strategy for the software to be utilised to enable asset management objectives has been defined as part of an overall Enterprise Architecture approach • The overall data model for the organisation has been defined and is used to assess the suitability of change projects • Software requirements capture considers the importance of good asset data • A clear approach to data mastering has been defined and supports ongoing business activities • Data migration activities are business led projects • Appropriate controls embedded to support e.g. data quality validation • Project methodologies specific data requirements and clearly specify business reporting requirements • Where data has to be duplicated across multiple systems, it is clear which system is master
15	Managing Change	• The organisation recognises change can arise in many forms and that change management processes include impact assessments for data and reporting • Change management is a business led activity supported by the IT department • There is close collaboration and good communications between the asset management and IT departments

Subject Specific Guidelines: Asset Management

System will be a core part of the operation of the organisation, will receive suitable high level input and support and will be reviewed on a periodic basis to maintain alignment with organisational objectives and the Asset Management System.

Governance of asset information will involve key senior stakeholders and will provide continual and effective oversight of both asset information itself and the processes and systems that create and utilise it. The governance body shall ensure that the Asset Information Strategy is developed and delivered. New requirements for information will be assessed as will appropriate responses to emerging issues.

The organisation's intent for long term management of its data and information production should be made visible in the creation and maintenance of an **Asset Information Strategy (AIS)**. This should clearly articulate the link between asset information and its role in achieving asset management objectives. It should define the target capability requirement so that change initiatives align to this common goal and be supportive in delivering a common vision. The AIS should demonstrate how the organisation's data can be exploited to optimise its value to the business.

Standards, Specifications and Requirements will have been developed to support the acquisition and management of data and in the production of information/reports. They will clearly state what information should be recorded, how it should be processed and stored and may include an Asset Data Dictionary to unambiguously define the way assets will be classified, named, related to each other and the attributes to be recorded for these assets.

At the heart of asset management information control, is the data which will be **actively managed throughout its lifecycle**, and will conform with the requirements gathering and definition as defined by Standard, Specifications and Requirements.
The lifecycle represents the
- acquisition of data
- its storage
- subsequent utilisation to support management reporting or analytics (for example)
- ongoing quality monitoring which should drive improvement initiatives if data quality is not sufficient
- archiving of data when no longer required on a regular basis
- subsequent deletion when data is confirmed to be no longer required to support asset management or any historic reporting

Business processes will have clearly defined data inputs and outputs to directly support the organisation's operation.
A good organisation will understand what data is critical for each business process operation and how this supports the associated measurements required to effectively manage those processes. Process owners have this level of understanding about their critical data, so they can help drive out improvements in its capture and maintenance. A good organisation will ensure appropriate business processes are in place to manage the data itself.

Software (or technology) takes various forms in supporting the asset management activity according to the scale of the operation e.g. ranging from a centralised EAM implementation to a simple spreadsheet based system. Good organisations will, for whichever form of technology support is in place, have appropriate controls for data in place to ensure data is validated and maintained to fit the rules relevant to the organisation. Wherever possible, the systems will automate data validation (for example) and that the overall systems components have data which is aligned (either automatically or manually) with clear definition of which systems is the master data source.

Monitoring in the context of asset information describes the ongoing activity of reviewing compliance with standards, specifications and requirements. A good organisation will ensure that the suite of standards etc. is actively maintained and the monitoring activity may highlight an individual training issue or perhaps where standards etc. need to be updated in line with new requirements.

Audit and Assurance activities describe two key management activities. Audit activity allows internal or external personnel to conduct independent scope reviews to ensure appropriate controls are in place to deliver the right quality of data to the business. External auditors may be engaged to assess PAS55/ISO 55000 compliance. Assurance describes the activity of verifying that the day to day management of data and information is effective. A good organisation will have a clearly defined and independent responsibility to review and monitor practices by perhaps reviewing a sample of certain data and/or processes. Assurance activities will interface with audit activities which will define a specific scope of work and 'deep dive' its review to understand conformance with standards and required practices.

Benchmarking will allow comparison of practices with equivalent (asset management) companies in order to determine comparative performance against other organisations. Benchmarking allows an organisation to identify potential improvements to be made or to justify internal KPIs to a regulator.

Managing Change is a frequent challenge in today's environment. There are a number of different drivers which impact the way data and information needs to be managed:
- Data requirements may change as a result of (for example) new maintenance policy implementation and perhaps more granular information is required. This will drive a consequent change in the data specifications and requirements;
- System or process change e.g. migration to a new system

3 Good Practice Approach for Asset Information Management

requires new data validations or processes change requiring data to be retained for longer;
- Organisational change might involve the change of geographical regions for maintenance or even the merger with another organisation.

The structure and culture of an **organisation** will impact on staff in successfully managing its data. These structural aspects will have been assessed when deciding on a centralised or federated approach to data management or report production with the appropriate level of controls to ensure optimum outcomes.
The people aspects of data management and the range of current and future required capabilities will have been assessed with appropriate activities put in place to promote good behaviours, at a more detailed level, the organisation will understand the depth and maturity of people skills and competences required to deliver on its required capabilities. There is an understanding that everyone has a role to play in data quality by, for example, highlighting an asset that has never appeared in a maintenance request.

All these components form a 'management system' for the organisation's data and information. Success requires a company to define the relevant balanced package of component management activities appropriate to its maturity of operation.

3.3 Key Messages

i. A structured approach to managing data will provide an effective and efficient means for an organisation to improve the benefits asset information provides to business activities
ii. Consider:
- Asset Information Strategy
- Standards, Specifications and Requirements for asset information
- Managing information throughout its lifecycle
- Monitoring, Audit and Benchmarking to review performance
- Ongoing governance, people and organisational factors
- Management of processes and systems
- Effective management of change
iii. Each organisation will need to assess how 'good practice' approaches may be relevant to their objectives
iv. 'Good practice' asset information management is aspired to by many organisations and will change over time

4 The Asset Information Management System

The organisation should establish an Asset Information Management System that will support the evolution of the whole Asset Management System and the developing information needs. Business processes and management activities should take place that continually improve and plan the development of asset information provision.

The approach taken throughout this SSG helps provide definition of the Asset Information Management System which, for brevity, is summarised in this Section.

Responsibilities should be assigned that define for example:
- The scope of each information set and how such information can be classified;
- How information is represented including the definition of, for example, drawing registers, asset registers, document control systems, change control systems, etc;
- Associated business rules for the collection, creation, acquisition, storage, use, transmittal, transformation, archival, retention and disposal of information;
- Roles and responsibilities for asset information management;
- Access rights and administration of access; and
- Version control of information.

In addition asset information needs will incorporate:
- The physical asset hierarchies – how assets are recorded within physical asset systems, defining assets that need to be identified as maintainable asset assemblies and sub-assemblies.
- Asset functional hierarchy – defines how physical assets comprise functional systems and functional sub-systems such as a signalling system.

4.1 Assess and manage risks

The organisation shall assess the business risks of not having particular asset information available at the right quality to the business and identify the criticality of different information sets. The organisation shall implement appropriate mitigation and control measures to ensure resilience and business continuity in relation to critical information.

4.2 Collect or acquire information

As part of business development in relation to improving the asset management system, the organisation will periodically commission special projects to collect or acquire new information to meet identified business needs. This may involve, for example:
- Bulk transfer of data from legacy data stores into new data stores;
- Asset surveys to gather and register assets that have previously not been recorded;
- Collect data on additional asset attributes that were previously not recorded; and
- Receive and store information on newly created assets as part of the project handover process.

4.3 Receive, use, transmit and transform information

The organisation will ensure that:
- Sufficient access is granted and made available to asset information; so that users are not constrained in carrying out their duties;
- Receive and record relevant information as part of ongoing business processes;
- Create information by combining a range of information sources; and
- When asset information is transformed; converted into other information forms or formats, that the process and outputs of the transformation are consistent and support the Asset Management System.

During information transmittal and receipt, both internal and external to the organization the organisation will ensure sufficient records are kept to determine:
- What information has been transmitted or received;
- The context under which the received information is to be used;
- The authorised user in the organisation who has sent or received the information;
- The information reference i.e. document number, drawing number etc.; and
- The time and date of the transmit or receipt transaction.

The organisation will ensure conventions are in place for asset information to be applied and used in the correct context.

4.4 Maintain information

The organisation will ensure that:
- Appropriate data quality KPIs are established to monitor the quality of asset information; and
- Sufficient care is applied by users in maintaining the quality of data in line with the data quality KPI requirements.

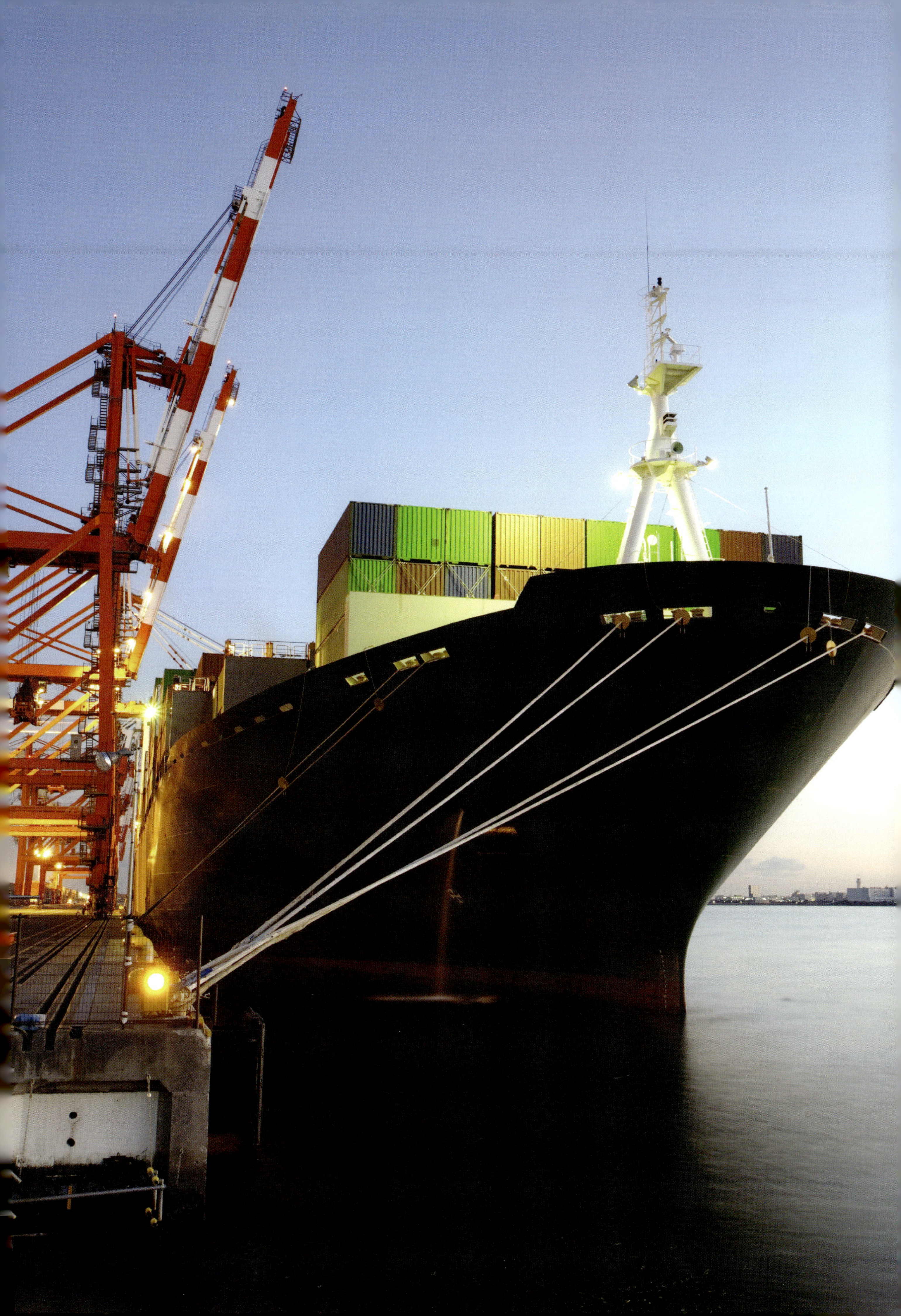

In relation to information security, the organization will ensure that:
- Information security policy and procedures that reflect business requirements are defined;
- The security environment is supported by management and understood and enforced across the user community, including contractors;
- There is an understanding of information security requirements and procedures are in place for risk assessment and risk management arising from breach of information security; and
- Effort is made to raise awareness of information security of all managers and users, and guidance exists in the form of information security policy and procedures.

The security management arrangement for asset information will include:
- The allocation of information security responsibilities;
- The authorisation process for information processing facilities;
- Contacts with law enforcement agencies, regulatory bodies and information service providers to ensure action can be taken quickly in the event of a security breach;
- The exchange of critical information being restricted to ensure confidential information of the organization is not passed to unauthorised persons (but without withholding access to it from business users who have a genuine need to use it); and
- Assessment of risks from third party access i.e. physical access e.g. to offices, computing facilities, filing cabinets and access to IT systems, databases and electronic documents.

4.5 Comply with legal and other requirements

The organisation will ensure that the asset information management system is compliant with applicable legal requirements. Consideration should be given to current and developing corporate policy, regional, national and international legislation.

4.6 Manage change

Manage significant changes to information definition, organisation of information, the Asset Information Management System and the information management process. Also manage the impact of changes to the Asset Management System.
Change should be managed using established change management procedures and should include, as a minimum:
- Consultation with stakeholders and major end users about the need and benefits of change and how this might impact them;
- Assess readiness for change and implement change at the appropriate time;
- Identify the impact of change and individuals impacted; and
- Train and support the individuals over a period of time until the change is embedded.

4.7 Archival, retention and disposal of information

The organisation will identify asset information records for archival and disposal from the Asset Information Management System.

Consideration shall be given to determining what action must be taken if the record is disposed of in error including determining how to re-create the record. In some instances the re-creation of some records is straightforward whilst in other cases it can be very expensive.

Approval should be granted for appropriate asset record archival or disposal.

Upon completion of record archive or record disposal / destruction, the status of the record should be updated. Appropriate policy and business rules should be defined for retention of information to meet legal and contractual requirements.

4.8 Periodic Review and improvement of the information management system

In conjunction with new requirements that have been described in the Asset Management Plans (or similar) the organisation will identify opportunities for improving the Asset Information Management System in support of the overall Asset Management System. The improvements will also be influenced by the planned improvements to the Asset Management System and / or other business drivers.

4.9 Key Messages

i. The Asset Information Management System is the overall approach to the management of asset information by an organisation and should align with and support its asset management system
ii. The Asset Information Management System defines:
   - The scope of asset information
   - How and where asset information is stored
   - Business rules for the collection, storage, usage and disposal of asset information
   - Roles, responsibilities and access rights to asset information
iii. Change is a typical part of any organisation, the asset information management system should define how it will support and accommodate organisational change
iv. The Asset Information Management System should be reviewed for effectiveness and support for organisational objectives with changes implemented where required

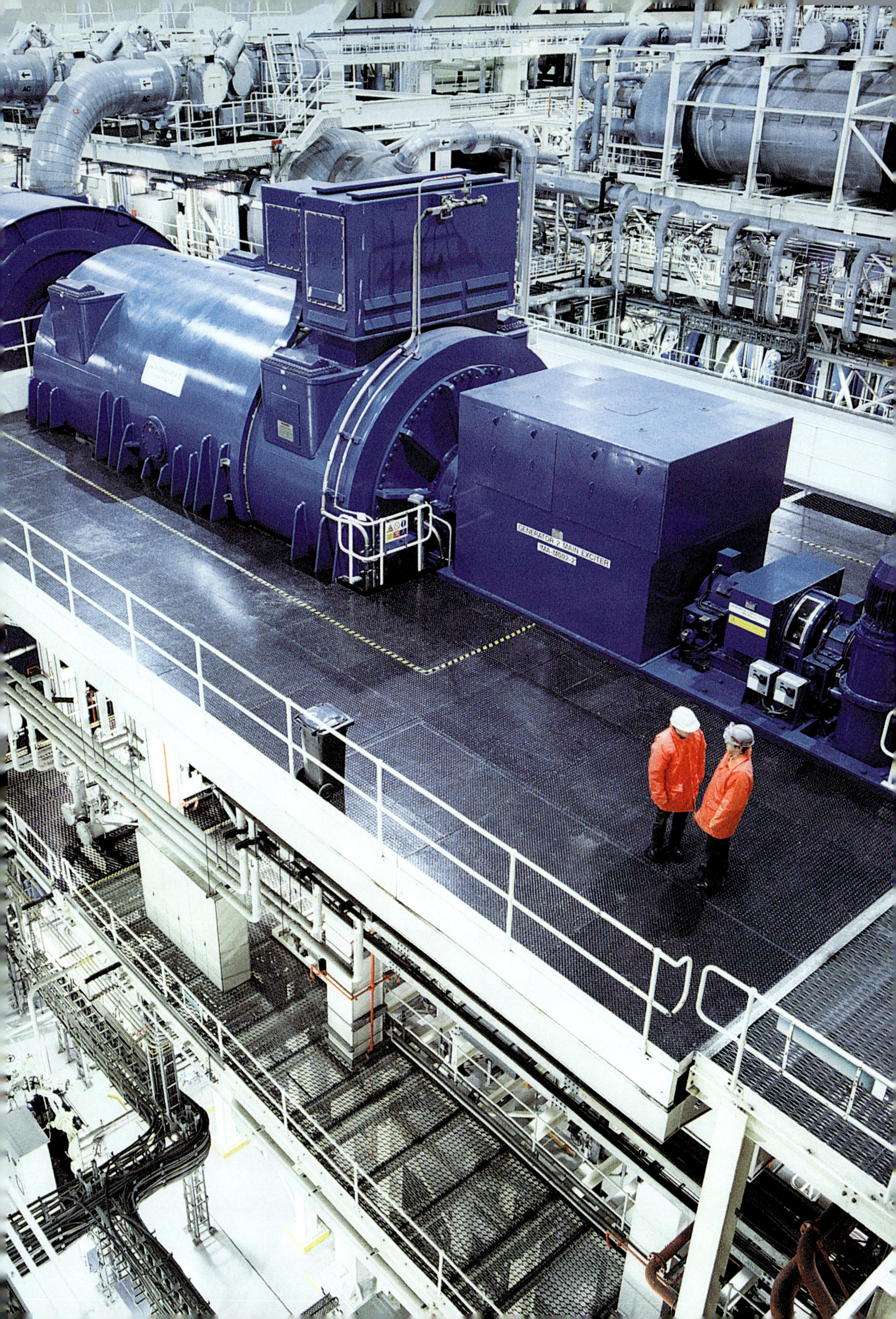
GENERATOR 2 MAIN EXCITER
1MA-M002-2

5 Governance

Information governance and support is how an organisation controls its information assets. It provides oversight of the standards, policies, processes, roles and responsibilities that direct overall management of an organisation's information. Governance provides the means to ensure that information is accurate, consistent, complete, available, and secure. Governance is one of the first enablers that should be put in place when looking to improve an organisations approach to asset information management.

One example definition of information governance is:

Information governance definition:

> *The act or process of leading, directing, controlling and assuring that information is managed effectively as an enterprise resource, including resolving information conflicts, across the enterprise.*
> **Larry English**

Governance should be an ongoing activity, however, what is often seen is that new projects, such as an information warehouse implementation project, have some element of senior executive support that helps provide the impetus to correct the known issues and results in the quality and availability of information increasing to acceptable levels. However, once the project ends and the senior executive support wanes then information practices can degrade. This is often due to Information Governance frameworks focused on finding a technological solution to a project rather than putting the people and processes in place that create and sustain the required information at the required quality.

Effective information governance is an important activity which ensures that an organisation can have confidence that the quality of its information will become and remain at the optimum level. In the absence of effective governance, improvements achieved through information cleansing or other information quality initiatives can all too easily be negated by other activities carried out without due regard to these improvement activities and an organisation's information quality needs. Information governance is a key feature of PAS 55 and should be incorporated alongside traditional business governance and IT governance structures.

A historical confusion about information quality is that the IT department or provider(s) traditionally regarded it as the responsibility of the business to improve, while the business regarded it as the responsibility of the IT department.

The reality is that everyone has to work together to achieve quality information within the organisation, and Information Governance provides the framework to make this possible. Information governance consists of three main elements, as illustrated in the following diagram:
- People, Roles and Responsibilities
- Management processes
- Knowledge of the Information required and available.

Processes	Meta-information	People
• Documented rules and requirements	• List of information assets	• Who manages information?
• AM and business policies and strategies	• Risk of information	• Information users
• Information standards	• Data storage landscape	• Information creators
• Quality system - local and corporate	• Asset management policies	• Business area sponsors
• Supply contract arrangements	• Business process maps	• IT Services
	• Quality of information	• Responsibilities

Figure 4 - The components of information governance

5.1 Governance Processes

5.1.1 Management arrangements

Governance processes need to be considered and addressed appropriately for each organisation. Some have legal or regulatory requirements for the frequency or quality of information updates. All organisations need to document their processes at some level.

The governance processes can be diverse, but must reflect that asset information is absolutely inseparable from the day to day asset management processes it supports and the software containing it. Any change in the process can have major impacts on the information contained within the information system.

5 Governance

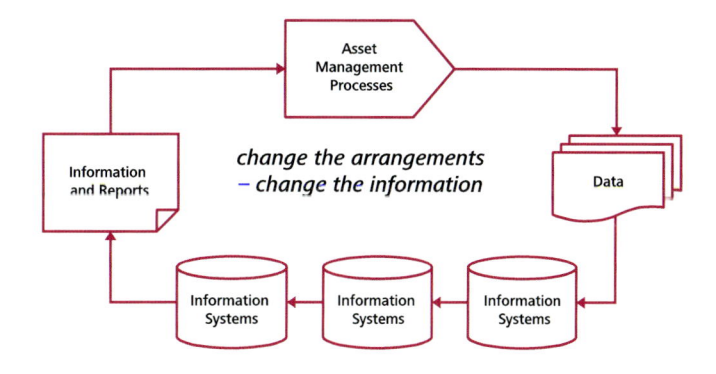

Figure 5 – Illustration of the impact of change on information processes

In practice, most organisations responsible for managing assets will already be doing some of the required governance functions, but probably in a piecemeal fashion with no overall framework or ownership. To establish a comprehensive governance structure, therefore, the first step should be to assess what is already in place, with the aim of then building on and around this to put the key missing elements in place.

In order for the governance to be successful it is essential that the existence and work of the information governance function is clearly publicised and that any party can easily raise information issues. It should also ensure that it is the sole influencing/decision making body for the parts of the organisation it covers. The governance processes need to oversee the establishment and agreement of information management strategies, standards, monitoring and plans.

5.1.2 Policy

An information policy should set out the aims of asset information for the organisation. This should include:
- Executive buy-in and sponsorship of the need for, and benefits from having, appropriate quality information, to ensure that sufficient resources are made available;
- Organisational objectives and KPIs for information quality, and policies for its management, based on quantified business benefits;
- Regulatory requirements, compliance with data protection legislation, and any other applicable external requirements;
- A framework for assessing organisational risk to identify potential impacts of information quality on business activities;
- The organisation's information security and access policy;

5.1.3 Strategy

There should be up-to-date strategic plans for the improvement and maintenance of information quality to set out the overview of how to achieve the information quality policy. This should
- Identify and define appropriate standards and rules for information quality (including definitions, formats, value ranges, thresholds, standards and interfaces to external systems);
- Define a disposal and archiving policy, for example - whether disposal of an asset should trigger archiving of its information, and for how long that must be held;
- Establish clear contractual requirements for the provision of information with all stakeholders. These should include provisions for both quality and timeliness; and
- Prioritise issues and information improvement initiatives. See Section 6 for more information on Asset Information Strategies

5.1.4 Implementation

The processes for implementing the information quality strategy should be set out in an implementation plan and clearly documented procedures. This should include:
- Transparent information reporting and clear processes to respond to emerging information quality issues;
- Easy processes for staff to report information corrections, changes and issues;
- Visibility of progress on initiatives to improve information quality and recognition and valuation of successful initiatives;
- Institute appropriate reporting mechanisms to monitor actual quality levels and measurement target performance within an organisation's performance management system;
- Establish escalation and conflict resolution procedures for information quality issues;
- Monitor all activities and projects which influence asset information quality, including those contracted out, and assess issues for resolution;
- Establish a information audit programme, including assurance of information quality review and assessment processes, and review its outputs;
- Keep everyone informed of information improvement initiatives and progress;
- Give full weight to information requirements in legislation in Business Policies and Strategies, although their application must be applied as part of the day to day processes;

> **Example:**
> *For a railway asset management organisation there are over 100 items of legislation requiring the maintenance of records. Whereas some of these can be applied generally, others can only be applied by the due diligence of local managers knowing what is applicable to them, working within the information management framework established by Company Policies and Plans.*

- Publishing information standards, to enable information to be shared across organisation boundaries by establishing requirements for creating similar information in different places, by different people. These can only progressively be implemented because old standards or a lack of consistency will already be built into the legacy information systems;

Subject Specific Guidelines: Asset Management

- Putting in place a level of documentation to provide sufficient visibility of arrangements and common concepts, business models and terminology that allows the Governance arrangements to work effectively; and
- Contract arrangements to protect the organisation's interests in information created by Suppliers with linkage to payment terms.

It is notoriously difficult to obtain information from Suppliers or Contractors, even when agreed contractually. Failure to include suitable clauses backed up by specific requirements for vital information will be virtually impossible to recover post contract award.

5.1.5 Review and improve
Management review of all aspects of asset information management is essential in order to ensure that the progress of information quality improvement activities meets business needs, that there are suitable resources deployed to improve quality and that timescales are being achieved.

Information processes need to be reassessed at regular intervals in order to determine whether the process is working as intended and also to evaluate whether the requirements and opportunities for information have changed.

This requires feedback, proportionate to the risk, on the condition of the information, performance of IT services and progress with information improvement projects.

Over time there will be changes to business objectives, the value of information quality can change and the available technology may be different. Periodic review of the Asset Information Strategy will be needed.

Agreed changes should be made to information policies and standards with the information quality requirements being reassessed.

As tracking of asset information activities is little different to other business activities, please refer to PAS 55 or ISO 55000 for more detailed descriptions of the Management Review activity.

It is also advisable to periodically review the overall effectiveness of governance processes in order to identify any areas requiring changes or improvement. Checks can include:
- The length of governance meetings and how constructive they are;
- The number of remedial actions being tracked by the governance group and the progress and timescales for resolution;
- Review of any identified information related issues which were not identified by the governance group in order to assess whether a change in approach may have allowed quicker identification and resolution of issues; and
- Review of the number and nature of information issues raised by wider business stakeholders in order to assess whether the governance processes are widely understood within an organisation.

5.1.6 Improvement
Activities to improve how information is collected, used and managed can be undertaken at most points in the information lifecycle. Possible areas for improvement include:
- Information collection technology and processes;
- Information collection activities and their interface into main data stores;
- The main data stores themselves (structure, information processing, access);
- Requirements for information to run the business daily;
- Asset information analysis and asset strategy development; and
- Ongoing requirements for data, information and knowledge to develop and monitor business strategy.

A holistic approach to improvement in systems and procedures is necessary to ensure that solutions are appropriate and that all interested parties are involved in finalising solutions.

It is important to allocate a sufficient level of resource to continue to maintain the improvement from an initiative to avoid deterioration in information quality when the implementation of the initiative finishes and the focus reverts to business as usual.

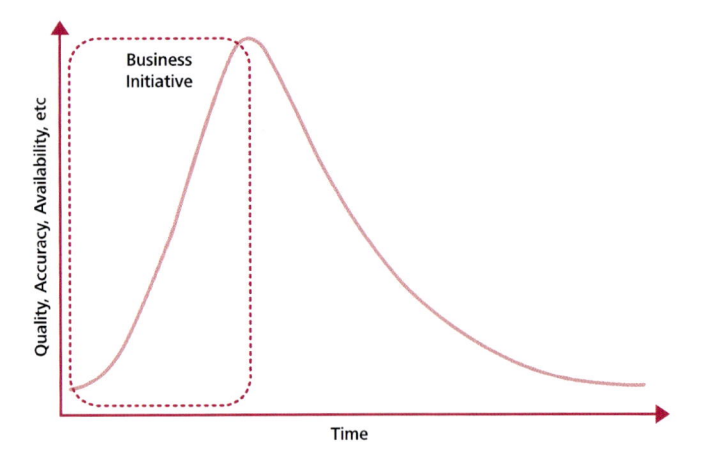

Failure to fully maintain and sustain information until it is no longer required can rapidly cause the actual and perceived value of systems to deteriorate.

5 Governance

5.2 Meta-information

It is important to collate and publish information about information quality governance so that all stakeholders have access to it. This should cover:
- Objectives, strategies and processes
- Catalogue of what information is available and who is responsible for it (see Section 7 for more information on Asset Data Dictionaries);
- Details of improvement initiatives planned, underway and completed.

5.2.1 Objectives, strategies and processes
- Ensure information quality objectives, measurement targets, policies, requirements and achievements are clearly documented, published and communicated to all relevant stakeholders;
- Establish and oversee an easily accessible repository of knowledge regarding the organisation's information;
- Ensure there are clearly defined responsibilities for updates to the knowledge repository with relevant sign-off processes;

5.2.2 Knowledge about Information

It has been estimated that an organisation applies over 20% of its effort to managing its information, however this effort is typically spread thinly and hard to marshal effectively. Understanding what information is held and required and its risk helps prioritise where it is necessary to apply effort to maintain information.

It is concerning that many managers don't realise what information could let them down, and how much the loss of information impacts on their business and personal performance until it is actually missing or the quality lets them down.

> *Individuals in an existing enterprise are already managing information to some extent with some understanding of its criticality, but may not have the support and resources to sufficiently mitigate the actual risk of loss.*
>
> *In this case when organisational change takes place without lack of formal information accountability and understanding of risks of loss of information, there is a possibility that such managers will be displaced and unable to carry out the maintenance arrangements which can result in rapid and irretrievable loss of vital information.*

For a new organisation or investment in a new major enterprise, the management arrangements described below can be established from day one to understand what information is required, its risk and the levels of control required to sustain the available information.

For an existing enterprise the level of resources available are likely to be much lower and probably require a pre-assessment to determine where to apply the effort most effectively.

5.3 Responsibilities

Responsibilities can be organised by the Lead Information Governor, who can assist in allocation of Information in data stores to owners. Agreement of the Business Area Sponsors with support of the Steering Group and Executive Sponsor is necessary as the allocation of information to owners becomes more difficult as software becomes more integrated. (This is because any particular manager identifies less with the content of data stores used by more than one department and may instinctively avoid taking on the inherent responsibilities and sometimes apparently intractable historic issues.)

> *Example:*
> *The Asset Register is used as the basis for all other asset management activity. If the asset register becomes out of date, the organisation will not know whether assets have been maintained - or whether the appropriate maintenance regime has been applied - with obvious safety consequences. It is therefore crucial that the asset manager owns and puts arrangements in place to ensure the register is up to date. Periodic audits are likely to be required. It is noted that the inherent likelihood of realising the risks of the Asset Register can be reduced by presenting assets on a (geographic or schematic) map, and asset attributes can be confirmed by checks carried out during maintenance. These will reduce the likelihood of undetected degradation and should reduce the level and cost of verification required.*
>
> *On the other hand information regarding work history and cost is certainly useful, and there are risks of the failure to record work done. Nevertheless, the most likely outcome of errors has the relatively low economic impact of rework, and over time the loss of a single record will be less and less significant – implying that this information requires a lower level of inherent verification.*

5.4 Key Messages

i. Information Governance is an essential activity for managing the quality of asset information
ii. Board-level sponsorship – governance must flow down through the whole organisation
iii. Strategy – vision of where the organisation wants to get to and how
iv. Legal and business requirements – some information quality requirements may be externally specified
v. Roles and Responsibilities – defined, staffed and appropriately motivated/incentivised

vi. Training/competence requirements – minimum competence requirements may be required for a measurement technique or to gain access to a restricted area
vii. Written policies, plans and procedures – appropriate in detail and style to ensure necessary quality
viii. Meta-information – need to understand information in depth to be able to improve and maintain it
ix. Governance processes – to drive information quality forwards and prevent any fallback
x. Continuous improvement – with incorporation of lessons learnt
xi. Ownership – Data ownership in isolation is considered by some to be an outmoded concept due to the many current and future users and uses of data. It is more effective to consider:
- Process ownership
- Data stewardship within a process
- System ownership across an organisation

xii. Establish an executive sponsor and a steering group to provide strategic direction to information governance. The membership of the steering group should cover all relevant organisational departments and functions;
xiii. Hold regular meetings of the steering group and circulate notes and actions from the meetings as widely as possible.
xiv. Define, agree and allocate information governance roles such as information owners and stewards, with clear Terms of Reference, performance objectives and incentives that ensure clear responsibilities and accountability for the quality of information;
xv. Clarify responsibilities for information quality throughout the asset lifecycle from acquisition to disposal; and
xvi. Foster a working culture that promotes information quality, for example by recognising and rewarding staff who take good care of information and encouraging one-team working between organisation and contractors' staff with regards to information and information quality matters.
xvii. Instigate training and communications to staff to ensure everyone knows the information governance policies and how they affect them

---

i. It is recognised that some organisations may not have a single source of truth for its register of assets. Visibility of a report making a realistic objective assessment of the economic and political risks of this scenario should raise the profile of this unsatisfactory situation.

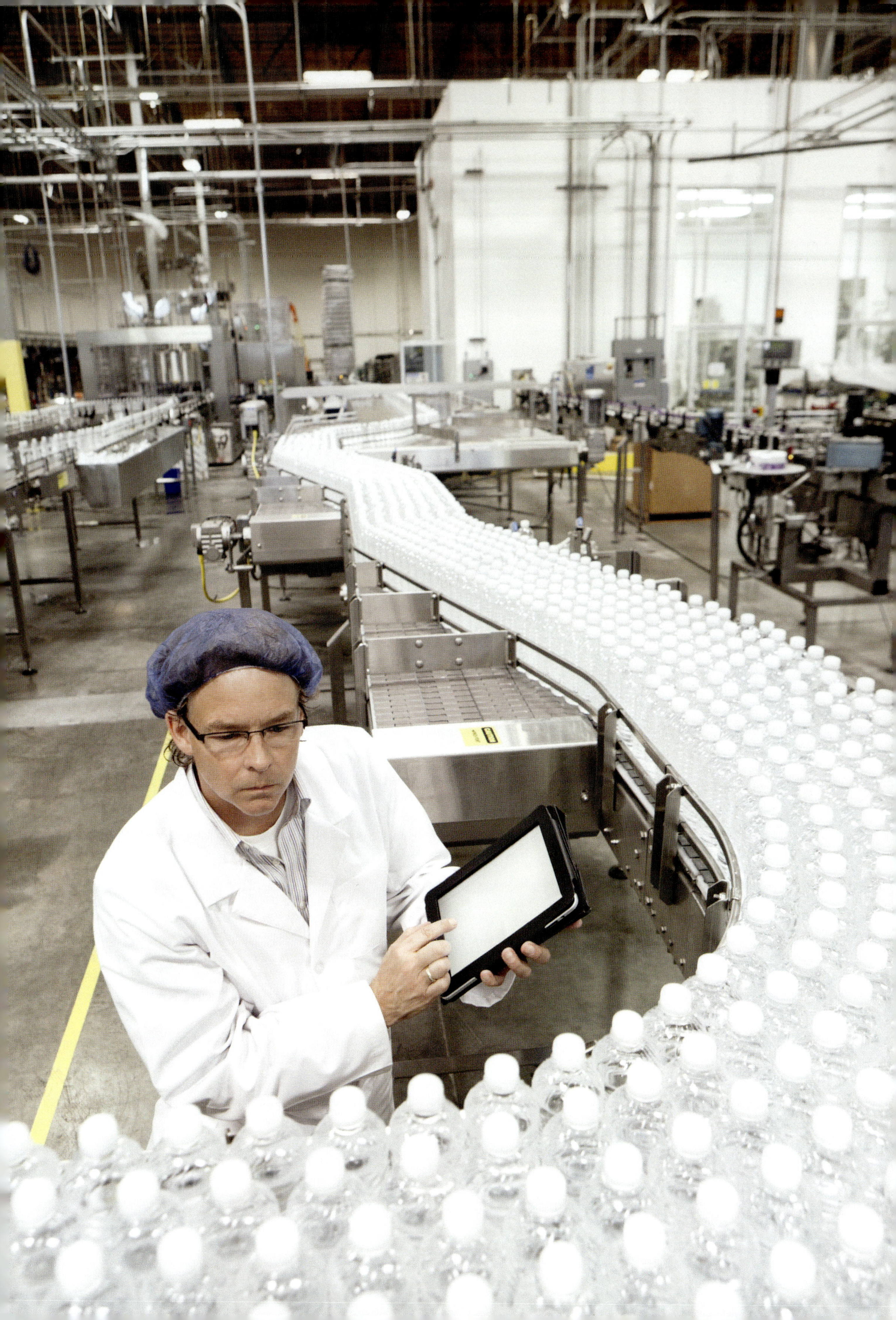

6 Asset information strategy

6.1 What is an Asset Information Strategy?

The discipline of "Asset Management" has within its scope the need to clearly understand its asset inventory, to understand where its assets are and to have the strategies/plans and policies established to optimise use of assets to maximise their value to their organisation.

An organisations data is no different - data is an asset in its own right, requiring a similar strategic approach and management activities which will vary in formality according to the size of the organisation.

An Asset Information Strategy will assess the current position and clearly articulate an "end state" or intent, in terms of business capability. It will make reference to
- The organisation's approach to data management detailing aspirations for business ownership of data, roles and responsibilities, its approach to data quality management, expectations around modelling requirements, and mandatory requirements for security;
- Defining the approach to the development, agreement and publishing of specific asset information requirements;
- The technology and software to be used to deliver the asset information strategy. This is likely to include a review of current software applications and their suitability, the approach to field devices for remote staff, interfaces and links to other organisational systems;
- Transforming data into trusted and accessible information i.e. through its Management Information delivery, the processes required for working to common metric definitions or sources for data;
- The vision for the organisation's data architecture and integration with both internal and external sources, and where international standards are to be adopted;
- Unstructured content management (e.g. design information, photographs, aerial survey records, documents);
- Business intelligence intent e.g. in building an analytical capability to discover new facts for the organisation's operation or policy formulation; and
- Requirements for the organisation's data governance i.e. where the data steering and decisions are made, arbitration when agreement cannot easily be gained, responding to issues and emerging requirements.

The Asset Information Strategy will also define how the organisation intends to move from its current position to the desired end state. This can include a statement of the required strategic changes, the timescales for change, responsibilities for delivery and the approach to funding these changes. Senior leadership support to the Asset Information Strategy is essential and is required to ensure alignment or integration with the core business strategies and objectives. Some organisations may choose to embed their Asset Information Strategy within their Asset Management Strategy – others may choose to develop it separately but clearly show alignment to the core business strategies, including Information Management/ Technology strategies.

The Asset Information Strategy also may be an overarching reference which defines a number of more detailed policies and standards. The strategy could support adoption of standards but then the decision has to be made as to which standards are most relevant and how to respond to changes to agreed standards. The subsequent policies and standards will then be the subject of separate review, definition and issue.

6.2 Why does an organisation need an Asset Information Strategy?

Whilst individual teams and departments may have successfully managed their part of the overall asset information landscape in the past, it is essential that a single, organisational wide approach is adopted in order to succeed in establishing, for example:
- What level of data management and accountability is required to fulfil asset management objectives;
- How to establish and maintain a 'single source of the truth' relating to assets;
- Plans for exploitation of data to drive the maximum value from it e.g. in moving from time based maintenance to condition based maintenance policies;
- Usage of messaging standards;
- Mandatory requirements for data due to legislation or regulation; and
- The overall capability required over, say, a 5 year period

Whilst departments and functions within an organisation may individually judge their data and information management to be adequate, a holistic approach should allow much greater benefits to be secured. For example, adopting common tools will be more cost effective, a shared approach to data quality monitoring will create a consistent set of quality metrics etc.

It is important that an organisation recognises the inefficiencies and difficulties meeting organisational objectives which can result

6 Asset Information Strategy

where there is no clear and agreed Asset Information Strategy. The symptoms arising can include 'dirty' data, duplication, data inconsistencies, lack of confidence in reporting, lack of accountability and a general perception that data is a liability rather than an asset.

6.3 Developing and Agreeing an Asset Information Strategy

The senior leadership team within an organisation need to be actively engaged in developing and agreeing the Asset Information Strategy – ideally with the support of an Executive Sponsor. This should be a business led activity with input from the Information Systems team and should not be an Information Systems initiative.

There are different approaches which might be relevant for different sizes of organisation. In some companies, the agreement of a few strategic principles could be sufficient. These then can provide the guiding framework for future projects and detailed standards etc. In other companies the Asset Information Strategy may become a managed reference document in its own right.

Four major reference points may help structure thinking to develop the strategy:
- People (skills and capabilities);
- Technology (systems and tools required to support implementation of strategy);
- Process (mechanisms for managing data and successfully creating value from it); and
- Data itself (what is critical to the organisation, what are the regulatory and legislative obligations).

See Section 14.2 for more information on Enterprise Architecture methodologies.

6.4 Implementing the Asset Information Strategy

In order to develop plans for the delivery of strategy, it is critical to understand the current state i.e. the current governance mechanisms, the current state of data quality, the existing standards and policies.

Having an agreed Asset Information Strategy and a clear view of 'current state' then allows an organisation to understand the scale of the task (the gap) to achieve its stated ambition. Funding constraints will usually require the organisation to agree a set of priorities and the sequence of improvements to be taken to start moving towards its goals. Typically, the data governance function will take the lead in ensuring that the options for implementation are assessed, prioritised and costed and will then submit them to the relevant organisational body for approval and funding.

Whilst the strategy is being implemented, the data governance function will typically take a leading role in monitoring delivery, responding to implementation challenges, assessing changes in priority and ensuring that stated benefits are delivered.

6.5 Updating the Asset Information Strategy

In order to remain relevant, an Asset Information Strategy needs to be managed as an active reference. It should be reviewed periodically, the frequency being dependent on the volume and significance of business change in terms of objectives, operation and where relevant, change in regulation, technology changes etc. If an Asset Information Strategy has been developed and agreed appropriately, then it should not be overly tied to a particular organisational configuration, therefore, an organisational restructuring should not require major revisions to the asset information strategy

6.6 Key Messages

i. It is imperative to engage the senior leadership team with an agreed executive sponsor in developing and agreeing the Asset Information Strategy
ii. The overall objective should be to:
   - State the current approach to asset information processes and systems
   - State the desired position; and
   - Define how and when this position will be achieved
iii. Ensure the Asset Information Strategy is current i.e. update in line with major business change
iv. The Asset Information Strategy should align with, and may be a part of, an organisations Asset Management Strategy
v. Data Management Strategy: Provides a long term optimised approach to management of the data derived from, and consistent with, the organisational strategic plan and the data management policy
vi. Data Management Policy: Provides a description of the principles and mandated requirements derived from, and consistent with, the organisational strategic plan, providing a framework for the development and implementation of data management strategy and the setting of the data management objectives
vii. Data management planning: Document specifying activities, resources, responsibilities and timescales for implementing the data management strategy and delivering data management objectives

Subject Specific Guidelines: Asset Management

7 Standards, specifications and requirements

It is important for organisations to have a clear and agreed view of the asset information required both for asset management purposes and where asset information supports other business processes. In most organisations, there will be existing data, which may conform to documented standards, however, this may be either significantly more than the organisation truly requires or significantly less. It is therefore essential that the overall approach is assessed, agreed and defined.

For the purposes of this document, the terms used are:
- Standards - Statements of how data should be captured, work activities undertaken etc. Standards can be internal, organisation specific documents or external, regulatory/legal requirements to be complied with
- Specifications - Statements of how data shall be gathered, processed and recorded
- Requirements - The organisational need that should be fulfilled, e.g. the collection of condition data for motors >22kW

N.B. These terms have overlap in meaning and may be used differently in different organisations

7.1 Why is Asset Information needed?

Before hoping to understand "what" information is needed, it is important to understand "why" it is needed. Without this it can become a case of "information for information's sake". Given the inherent costs associated with information management, the reasons for having it should be well understood and quantified in terms of the value it adds to the organisation.

Information should be needed to support one or more defined business process, for example:
- Strategic – at a level to support a "Strategic Objectives", for example measures of corporate business drivers of safety, performance, environment, etc.
- Tactical – to support "Technical Policy Development" (e.g. data and information to support analysis of maintenance intervention policies)
- Operational – to deliver asset related work through the efficient scheduling of resources (e.g. location data to perform work at site)

The use of Asset Information is not restricted to Asset Management practitioners; the users are wide and varied, and in some cases are not necessarily located directly within the organisation. Depending upon the audience and the questions to be answered, the information needs differ in their purpose and required level of detail. However there is a common vested interest in its availability and correctness.

Figure 6 - Users of asset information

The range of users of Asset Information is illustrated by Figure 6. With the exception of the external stakeholders such as Regulators and Customers, the degree of information granularity needed by each group increases the further from the strategic centre. Importantly, information should be consistent across all levels of the organisation. For example, bottom up operational information should align with the top down strategic information, and business reporting should be based on a "single source of truth". It is important to note that the level of detail required for a particular information requirement is likely to vary between different data users.

Ideally all requestors of information should be able to justify the reason for their information request and articulate the business process that it supports. This exercise requires business processes to be understood along with the business unit responsibilities, interactions and handoffs - process information requirements can then be gathered and collated. A method for drawing out

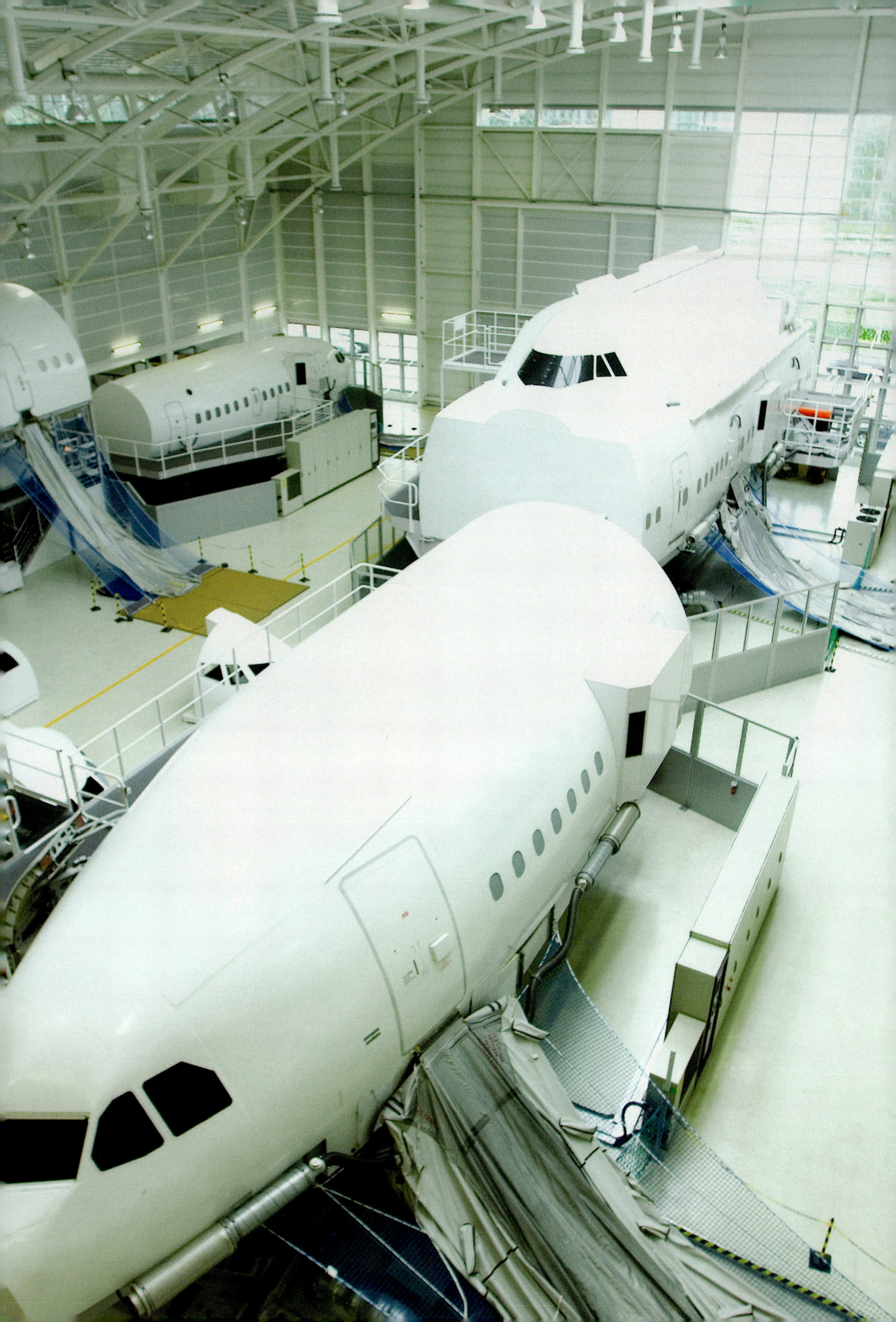

requirements is to ask what questions need to be answered at each point of a process.

This exercise should be undertaken with key representatives who are aware of the business context of the information and the importance it has to their areas. A common mistake is that the forum convened for collecting requirements is ill conceived or not represented correctly. For example, an operational manager may not have the knowledge to define the information requirements needed to set the corporate strategy of an organisation, which should be understood by strategic planning analysts. The structure for capturing, reviewing and approving requirements should be well defined and communicated to the relevant parties.

7.2 What Asset Information is needed?

Once the business needs are understood, more detailed requirements can be defined. The initial requestor should be well placed to confirm what is specifically required. By its very nature, Asset Management requires data from a variety of disparate sources within an organisation; typical examples include:

- Physical asset data – what assets are owned/operated and what are their technical characteristics?
- Location and spatial links - where is the asset and how does it relate to other assets?
- Relationships – how are assets related to each other?
- Work Management Data - what work has been /will be performed on this asset?
- Performance Data – how does this asset contribute to serviceability targets?
- Condition Data – what is the residual life of the asset?
- Cost Data - how much does the asset cost to buy and operate?

In addition to the information above, other details need to be understood:
- Frequency of use – how often the information is needed in order to support the business process;
- Type – the type of data required for a business purpose;
- Specific attributes and units – e.g. the age of an asset in years; and
- Precision and accuracy – e.g. how exact measurements need to be and to what extent can inaccuracy be tolerated.

The activities of determining the information needs of an organisation should form part of its overall Enterprise Architecture/Information Architecture approach (see Section 13.2). The outputs of such work often include a data model that comprises the information to be held by the business throughout the asset's life. This should include:

- Attribute content;
- Interrelationships with other information;
- Measurement units;
- Format.

It is important that the Enterprise Architecture approach supports the development of a single, holistic approach to data definition and tries to prevent the creation of separate data silos that cannot be combined easily in future. Additionally, as data acquisition activities may take significant time to complete, it may be prudent to define data requirements based upon what the organisation will feasibly require in future, rather than solely what is required at present.

7.3 Considerations during requirements gathering

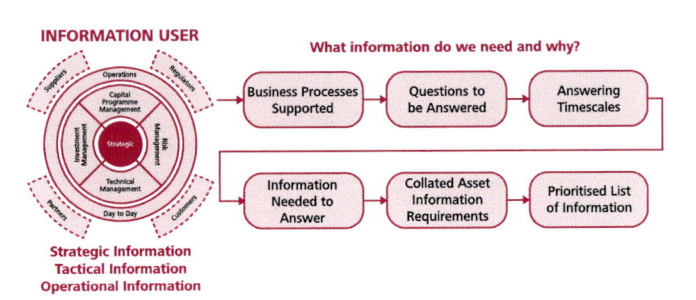

Figure 7 - Gathering requirements

Due to the nature of a requirements definition process, encouragement should be made for short digressions regarding the information availability, associated cost and benefit of having the information, data governance limitations and technological considerations regarding the Information Technology solutions.

Having these conversations at this stage will result in asset information requirements that are realistic and achievable, as indicative costs and benefits will determine those requirements that should not be pursued. Conversely, requirements should not be overly restricted by existing technology solutions or cost considerations at this stage as these can change based on solution options.

7.4 The Asset Data Dictionary

The collected standards, specifications and requirements for asset information are frequently described as the Asset Data Dictionary for the organisation. This specifies a set of common characteristics for Asset data, which must be adhered to for the data to be considered fit for a specific purpose. An Asset Data Dictionary typically specifies:

- How assets relate to each other, which is typically in the form of a hierarchy. Different levels in the hierarchy will be defined, similarly, the level of detail at which the asset hierarchy stops is specified;

7 Standards, specifications and requirements

- How assets are classified, i.e. grouped by generic type. Different classification schemas can be adopted, such as Uniclass or Omniclass, however, it is likely that whichever approach is selected, it will need to be customised to suit the needs of the organisation;
- How assets are defined by function, i.e. generic purpose or duty;
- Where assets are located;
- The attributes relevant to a particular class of asset and the definition of these attributes; and
- Rules for how to define assets in different circumstances.

This in essence defines the data quality rules for the organisation. Therefore, the need to adopt the Asset Data Dictionary is crucial to data quality as is the need to ensure that the adoption of these standards is pervasive across the organisation internally, and shared and agreed with external partners (i.e. the concept of the extended enterprise information).

The process of agreeing data standards is essential to ensuring ongoing maintenance of data quality but can also be applied to identify business process areas that are below standard data quality. Data standards could include defined ranges for numerical data, defined lengths of text strings and mandatory data fields. Improvements in data quality can be achieved by testing compliance with such standards (using electronic processes) and developing proposals to address shortfalls, whether this is by inference or simple reformatting of data.

Although data standards can easily be set out, asset owners may need to consider mechanisms to ensure that the providers of data (internal and external) actually do provide asset data to the agreed standards and timescales, for example, on completion of a capital project. This could be achieved through holding retention monies, or including the data standards as part of performance metrics. Additionally, there needs to be consideration of the time and work required to adapt systems and applications to comply with new or changed standards.

It is important that the level of detail being requested is appropriate and affordable to the organisation. The information requirements of one team may be onerous for data gatherers and, if analysed on a cost/benefit basis, may not be unattractive for the organisation to pursue, unless there are other drivers e.g. regulation or legislation, requiring that information. In case of doubt of disagreement, the governance function should arbitrate and support development of standards suitable for the organisation.

Another form of standards to be considered are the formatting standards to be adopted for communication between internal and external parties. It is not efficient to adopt different formatting standards for each type of communications link for each supplier/customer. A consistent formatting standard will allow simplified maintenance and be more cost efficient than multiple variants of interface formats.

Ideally the data dictionary should be published on an Intranet portal, or similar, where the quality information can be updated by the data stewards as and when necessary, and where anybody in the organisation can look this up when they need information either relating to requirements for information provision or to understand the meaning of data.

7.5 Key Messages

i. The taxonomy of your assets provides a common language and defines the structures, relationships and classification of assets and how they support their respective business processes
ii. There are different audiences for Asset Information whose requirements will differ in precision, granularity and complexity
iii. All requirements should have a purpose and reason and ideally support a business process - answering why information is needed is the first step in building a business case
iv. A common information thread should be seen through strategic to tactical to operational information
v. Information usually serves more than one stakeholder each of whom may have different time horizons
vi. The process of defining Asset Information requirements needs sufficient and carefully chosen representatives to be made available
vii. Consideration will be needed of how your agreed requirements differ from the theoretically perfect that you would desire based on cost/benefit, available resources and technology limitations
viii. Asset information requirements should not be 'tied in' to a particular organisational configuration as these will change over time
ix. Requirements should be available to all relevant stakeholders perhaps through the publication of an Asset Data Dictionary
x. Note that the scale of the task of developing and agreeing requirements should not be under-estimated

8 Information lifecycle

Information of any kind has its own life cycle. Data is created, used, maintained and archived when no longer pertinent. For asset information, this lifecycle may mirror the lifecycle of the physical assets the information describes and so it can be useful to consider such information in a similar way to an asset when embarking on information activities.

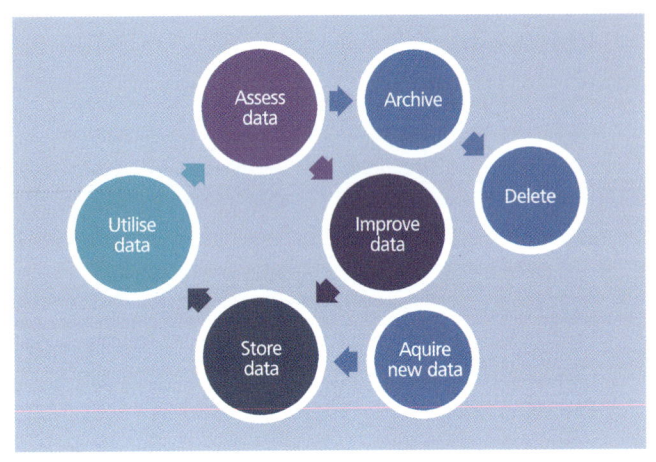

Figure 8 - Information lifecycle

Organisations typically will not be in a 'green field' situation and will have existing data, perhaps of uncertain quality, that supports business activities. This SSG assumes that this will be the situation and that one of your first actions will be to assess the current state of your data. This may indicate a requirement to improve existing data which will then need to be stored effectively. Data can now be utilised to provide business benefit which completes the basic information lifecycle. New data may be acquired either through the creation or acquisition of new assets or by gathering more data on those assets already recorded. An outcome of assessing data may be a need to archive data which will move it from 'live' storage, but will allow the data to be retrieved again, if required. Finally, when there is no further need for data, it can be considered for deletion.

Organisations should ensure that they understand the information life cycle and utilise effective management of the quality of asset information. This will help reduce the costs and maximise the benefits over time. This can be illustrated by adapting a famous quotation about the cost of quality:

> **Data Quality is free:**
> *Data Quality is free. It's not a gift, but it's free. What costs money are the unquality things – all the actions that involve not getting data quality right the first time and all the actions to correct these data quality issues.*
> **Adapted from "Quality is Free", Philip B.Crosby**

This indicates that the action of getting Data Quality 'right' will minimise the cost portion of the information lifecycle and maximise the benefits that can be derived from data.

8.1 Assess data

There can be many reasons why information degrades, for example:
- Not being kept up to date;
- Not being sufficiently accurate or complete (when updated);
- Data on the same asset in different systems being referenced differently or becoming inconsistent over time;
- Minor or gross errors that may or may not be obvious, including the right data entered into the wrong place;
- Inconsistent terminology or descriptions leading to loss or incomplete reports when searching or aggregating data;
- Software unavailability or catastrophic failure;
- Failures of software or interfaces to deal correctly with all types and orderings of data change lifecycles;
- Poorly executed data migrations when new software applications are implemented; and
- Business or technology changes render the information or data difficult to access or create disconnects between new and legacy software and data formats.

Degraded information can lead to a range of negative outcomes, either directly or indirectly, such as:
- Asset failure due to gaps in work planning and reporting;
- Safety incidents due to risks not being identified, reported or communicated appropriately;
- Poor asset performance due to a lack of information and hence understanding of asset behaviour;
- Non-compliance with statutory or safety requirements;
- Commercial inefficiencies or claims due to information disconnects across the supply chain and through regulatory interfaces; and
- Inefficient whole life cost management of assets due to a lack of information to enable robust asset management decision making.

Before information can be published or used to make Asset Management decisions, the availability and quality needs to be understood. Making decision makers aware of what information

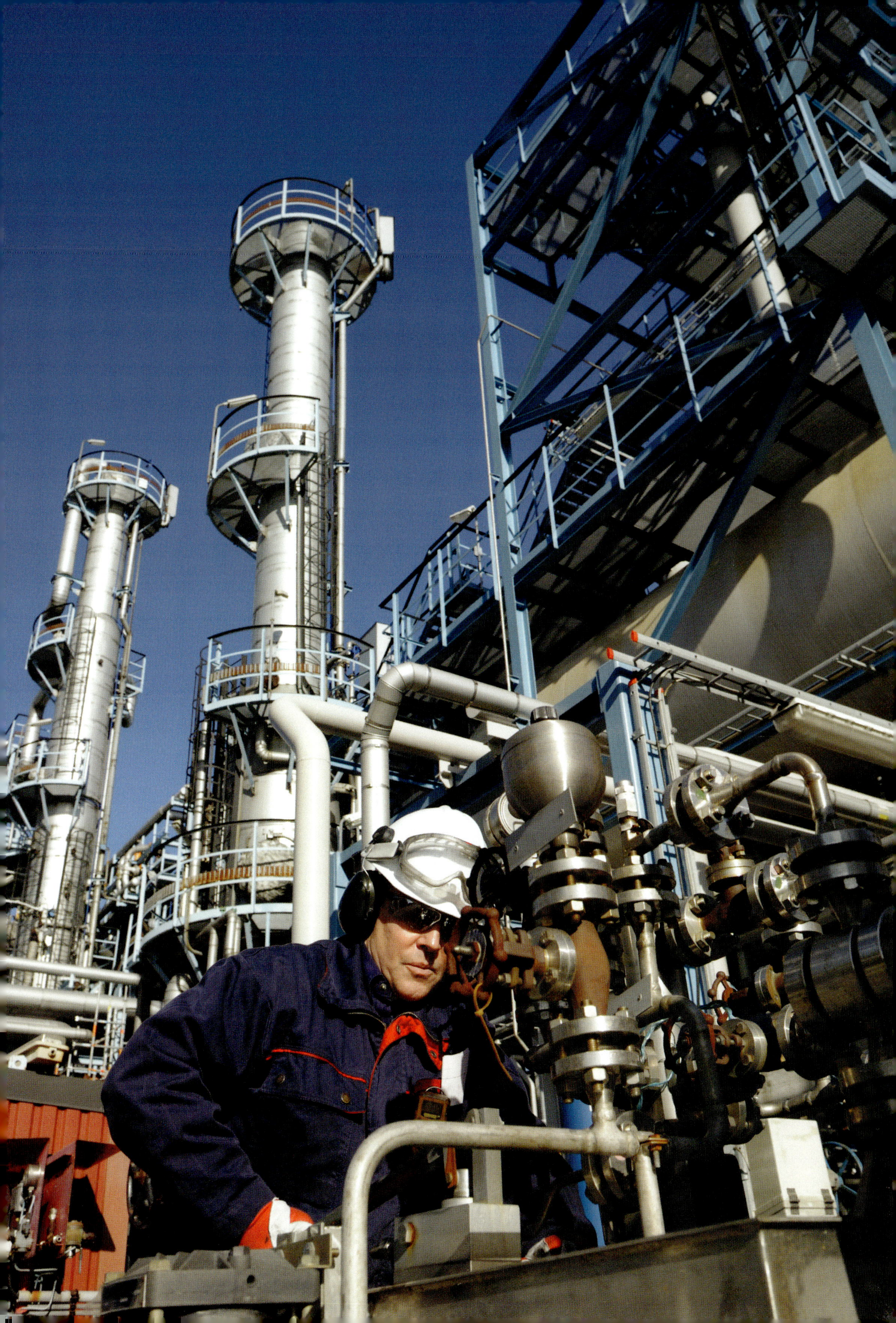

is available and the quality of this information allows properly informed decisions to be made. Without this being known there will be limited confidence in the accuracy and effectiveness of the decision.

8.1.1 Current State Analysis

When trying to understand what information is held and what is important to the organisation, it is useful to conduct a 'current state analysis'. The analysis of the current position will drive more detailed questions; the responses help determine the business requirements for improvement of its data.

Current State Analysis approaches are addressed as separate techniques as follows and presuppose that the business has a sound understanding of the quality of data required by the business (which should be recorded in the Asset Data Dictionary, see Section 5.4):

- What information do we have and why do we need it?
- What quality of data is sufficient for the organisation and is there a level of uncertainty we can live with?
- What toolbox options exist for assessing existing data quality (which can include specialist data profiling tools)?
- Do we manage our data in the right place or point in the process?
- Do we measure at the right point?
- Is data accessibility appropriate?
- Are there benchmarks available which help to inform current position?
- What is the organisational audit approach to current state assessment?
- How is the current state assessment visualised?
- How is data verification applied to the current state assessment?
- Are compliance and assurance checks undertaken?

8.1.2 Data quality assessment

If data was described as 'poor', this could be interpreted in different ways; similarly, if someone describes the weather as poor, they need to clarify whether it is too hot/cold, dry/wet, still/windy etc. When assessing data quality, a number of common 'dimensions' of data quality are typically used:

- Accuracy – does the data correctly represent the asset it relates to?
- Validity – is the data stored in the correct format?
- Completeness – Are all assets and required attributes populated?
- Consistency – Does the same asset have the same identifier across data sets?
- Uniqueness – Is each asset recorded once and once only?
- Timeliness – What is the time delay between a change to an asset and the corresponding data change?

Any assessment of data quality should be made against the data requirements detailed in the Asset Data Dictionary which should form the basis of an organisations data quality rules. A variety of methods can be used to measure or assess data quality against these data quality rules; however, the techniques that can be used vary depending on the data quality dimension.

Data profiling tools, SQL reports and spreadsheet analysis can be utilised to assess validity, completeness, consistency and uniqueness. Whereas, to check accuracy will typically involve someone assessing a data entry against the actual asset it represents, possibly by visiting that asset. Timeliness may need to be checked by comparing data update dates against the appropriate event that triggered the change.

Data profiling tools are specialist software/analytical tools that include a number of standard checks that can be made on data sets and often provide ways to utilise and enhance existing data quality rules. Although some tools can appear expensive to acquire/utilise, they can significantly reduce the effort needed to assess the quality of a data set.

Completeness
- Are all assets and attributes recorded?

Consistency
- Can we match the asset data stores?

Uniqueness
- Is there a single view of the asset?

Validity
- Does the data match the rules?

Accuracy
- Does the data reflect the asset

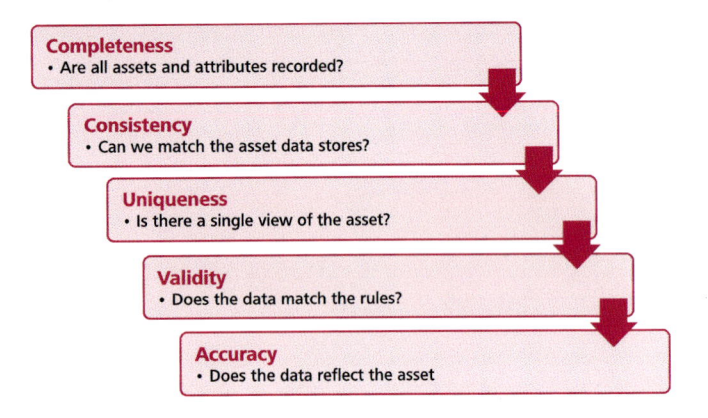

Figure 9 - Application of data quality dimensions

8.1.3 Information availability and requirements

Current state assessment must analyse current data availability, future data needs and understand why the organisation needs specific sets of data. This will aid determination of the quality requirements for that data. If no clear purpose can be confirmed, then it is likely that an organisation maintains data unnecessarily.

A key question in this area should be to address the sourcing of data and challenge whether data is sourced from the most relevant and accurate source.

It is necessary to evaluate the business costs and risks of not having data or of having poor quality data as well as the costs of obtaining data of differing quality. What correctness, completeness, consistency etc. (see Section 7.1.2) are needed to manage the risk of data causing poor operation or business decisions? Figure 7 (above) illustrates items that should be considered when managing data quality and provides a checklist for assessing the requirements at each level in the business:

8 Information Lifecycle

- Risk management;
- Benefits of managing Data Quality;
- Value of data quality;
- Cost of uncertainty – risk avoidance – inability to make decisions;
- Avoidance costs;
- Legal and regulatory requirements;
- Fit for purpose; and
- Archiving and deletion requirements.

> *Lessons from experience:*
> *It is useful to remember the 80/20 Rule where you can collect the first 80% of data with 20% of your resources however the last 20% of data is likely to consume the remaining 80% of your resources.*

For one organisation collecting data to complete material masters on an enterprise database implementation, this sort of consideration led to the stopping of the data collection process at the 95% complete mark. At some point it is no longer worthwhile getting better data, a conscious decision to stop collecting data may need to be recorded on the risk register and managed using risk management processes.

Based on the requirements considered in Section 7, the availability of the information that is needed can be assessed. With the correct Information Governance structure in place this assessment can look to the supporting standards and policies for guidance.

The answer to availability of data is often not a simple "yes" or "no". For example the information may be available but not in the correct format or the information not available but there is an alternative that could be used. This activity can be iterative, with the initial requestors of information being involved to confirm their willingness to compromise or modify requirements.

The following are the types of issues that will be uncovered during this activity:

- Entity issues – data about these things is not captured;
- Attribute issues – a piece of data, about this thing is not captured;
- History issues – data is currently captured but there is no history;
- Relationship issues – data is captured but there is no relationship (or the wrong relationship) between data sets; and
- Accessibility issues – data is available in the organisation but difficult to access.

For information that is not available a business case will need to be developed. For information that exists in one form or another, it needs to be assessed in terms of its quality before it can be used for decision making.

8.1.4 Data confidence grades

Confidence grades may be considered as part of quantifying current state assessments. Confidence grading explores the link between assessing the data and the confidence you can place on the data, thereby determining the confidence the business can place on the decisions made which use the data. This is with particular reference to regulated businesses where regulatory submissions may be accompanied by confidence grading assessments.

For large entities, the realisation that their asset information may always have anomalies and deficiencies is a sign of asset maturity. Should confidence in data be so low that decisions are not evidence based, but on an 'expert judgement' basis, then the decision process will require clear qualification and revisiting when data deficiencies are addressed.

The annotation of data reports with confidence grades is a common feature in regulatory reporting and provides a useful style of metadata for report reliability. One example of regulatory confidence grades used in the water industry in the UK is made up of two components – an alpha character for reliability A-D; and a numeric grade to denote bands of accuracy percentage. Inventories that require any extrapolation within the key attributes or an infill exercise for full reporting affect percentage accuracies and overall reliability of reports.

Confidence grades can be readily applied to a population of data entities where the accuracy component will be based on sample surveys. Confidence flags on each data record or data parameter are harder to assess but can be used to categorise the source of data, from which the reliability of data may be tracked. Parameters from site survey or inspection have high reliability grades while data based historic drawings or inferred have lower reliability grades. Similar flags should be integral to data collection and validation procedures to track progress and tag anomalies for further investigation.

Confidence grades can refer to individual parameters, data records or whole inventories. Confidence grade systems may differ at acquisition, refinement, inventory reporting and derived reports from data sets. Governance procedures should define how confidence grades are applied through inventories and processes. Data initiatives should consider confidence flags to monitor progress.

The source of the data may be from a system that contains rigorous controls with close monitoring of the quality or a locally maintained spreadsheet. The confidence in the locally maintained data would be low compared to that of the system.

Where a complete dataset is not available it is permissible to extrapolate data based on a small amount of reliable data, however this could lower the confidence placed on the inferred data.

> **Case Study – Water Infrastructure Acquisition**
> *The uptake of paper records onto a Geographic Information System collects not only location, but also the asset attributes diameter, material and date installed among other parameters. Attribute population relies on lengths on drawings being labelled with this data. Paper records were of varying age and quality with sparse labelling and the intersections between old and renovated pipe unclear. Confidence flags allowed attribute reliability of principle parameters to be recorded within the uptake process, with uncertainty of pipe details not delaying record uptake. A decade after record uptake, the availability of infill of parameters by automated GIS processes suggested an expansion of the confidence flags to clarify parameter infill basis. Burst repair returns containing diameters of fittings installed offers scope to verify or downgrade confidence flags during day to day interaction with assets to gain confidence in the reliability of inventory records.*

8.1.5 Consequences and Impact of Information Gaps

Asset Information users feed into this discussion as they are best placed to detail the business impact and consequence if nothing is done to correct issues, or to deal with the identified root causes. For asset management the following needs to be assessed:

- The relative contribution of gaps to the overall asset information quality, and what the overall business impact of the gap is;
- Future risk (e.g. consequence of individuals retiring, or a contractor changing), including an assessment of the resilience of the current and future system; and
- Consequences in terms of direct and indirect cost, and impact on other core business values such as reputation and regulatory confidence.

8.2 Improve data

There are a range of possible approaches to improving data quality within an organisation. These can include step change initiatives to achieve data quality targets over a defined period and changes to business processes or corporate systems to achieve targets over a longer term. Development of an overall strategy is likely to integrate initiatives across this spectrum to improve the quality of data being collected or to provide additional system functionality. Such approaches should seek to demonstrate their success by embedding data quality improvement into sustainable business-as-usual processes.

Improvements to data quality should have clear targets so that the objective and the end point are understood by all participants. Data quality within an organisation will change both as a result of planned activities and through day-to-day business activities. Therefore improvement projects must be designed to meet broader business improvement initiatives, like ISO55000/PAS 55 compliance.

A useful approach to the effective management of asset information quality is to treat data quality problems as symptoms of other, underlying root causes. While strategic improvement activities should be focused on addressing these root causes, the time to deliver these changes may be beyond the timescale available. Therefore a solutions' benefits can be operational (short-term); tactical (medium term) or strategic (longer term), each with differing cost and benefit profiles.

Whichever approach is appropriate to a particular situation, it is essential that activities to identify the most suitable solution are undertaken. A range of possible approaches are discussed below.

8.2.1 Data Quality Frameworks

A valid approach to improve data quality can be to act on each data quality issue as it is identified. Data Quality Frameworks can add structure to such an approach and provide a suitable escalation route for more significant issues. The key feature for Data Quality Frameworks is the focus on allowing data to be improved over time and to focus initial step-change exercises on high priority or time critical data.

The governance of a Data Quality Framework typically includes:
- Workflow systems to allow issues to be directed to the appropriate data stewards, and progress on resolution to be monitored;
- Assessment of the root cause of a particular issue;
- Recording of the extent, impact and cost implications of a particular issue. This includes both the cost of resolution and the costs that may be incurred if an issue is not addressed;
- Recording of decisions about whether to address a data issue, and if so, what the agreed resolution method is;
- Confidence flags on parameters to monitor progress;
- A dashboard facility to allow the current status and reliability metrics to be monitored by the data owners and other senior managers;
- Discovery/profiling analysis within routine audit procedures;
- Automated verification capabilities within acquisition processes based on business rule bases;
- Facilities to allow automatic detection and reporting of anomalies; and
- Escalation routes to ensure that emerging wide scale data quality issues are addressed appropriately.

The number of occurrences of each data issue, and the practicable options available for correcting the data, are often key factors when assessing the costs of resolution. For example for small numbers of errors, it is usually easiest to get these corrected manually in their master system(s) by the relevant data

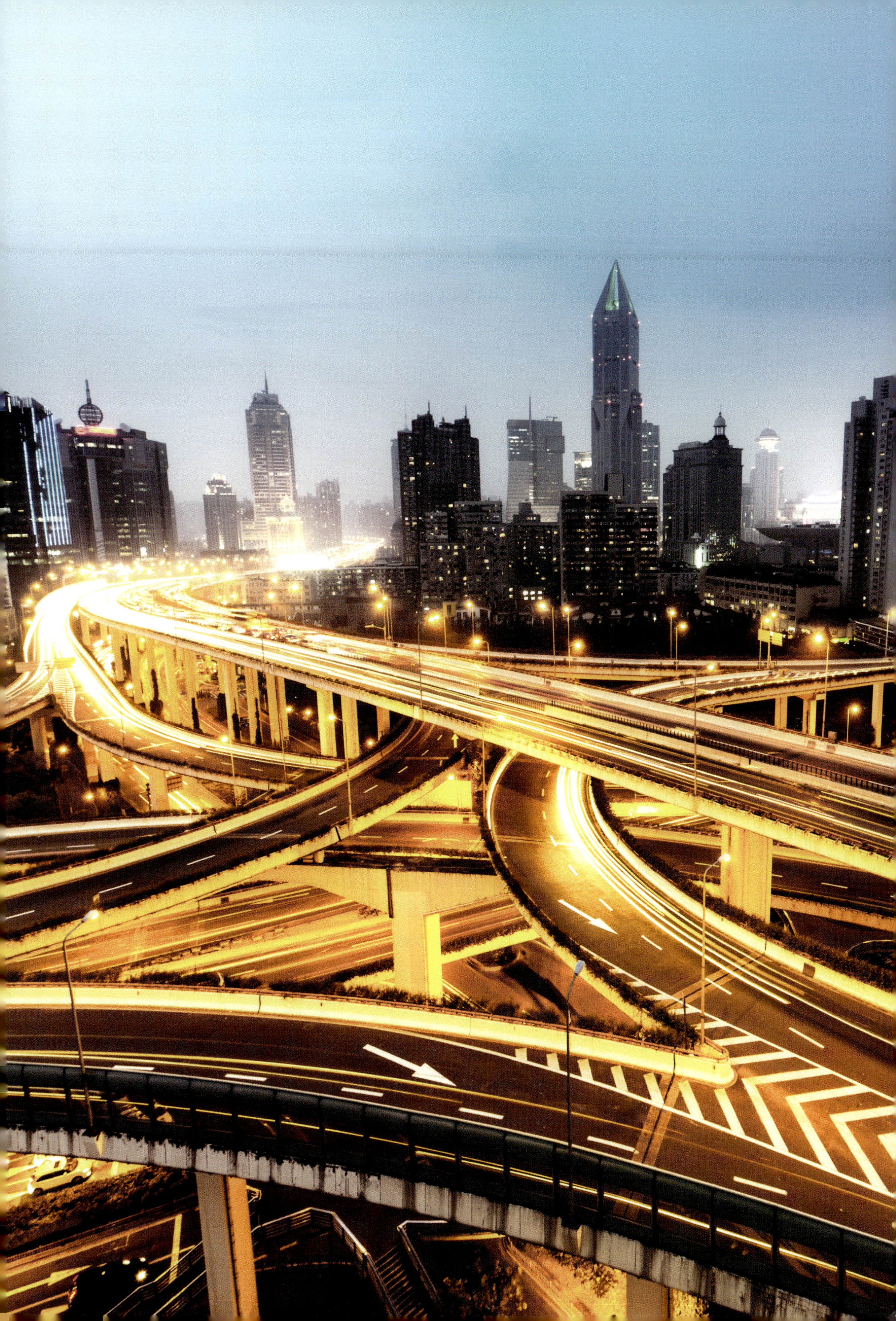

team(s), providing the use interface(s) will allow this. For larger numbers, explore whether all or most instances could be corrected automatically by cross-referencing to other related data, with assistance from the IT support teams if required. If a Master Data Management system has been deployed (see Section 14.4.1), it may also be possible to configure this to automatically propose corrections based on inference rules).

8.2.2 Root cause analysis

If monitoring indicates poor or unexpected performance, then a root cause analysis can be carried out to understand the underlying causes of information quality issues. Areas to investigate may include the following:

People
- Is there a culture of developing 'local solutions' at the expense of corporate solutions
- Do staff provide all required data updates and corrections?
- Do people understand the importance of the data?
- What are the incentives to get it right?
- Is it easier to do the right thing?
- Can we measure performance for data roles?

Processes
- Are the information requirements and outputs of a delivery process clearly defined
- Have the requirements been communicated / specified?
- Is the process for capturing the data documented & understood?
- Is the process 'straight forward' to follow?
- Is the process open to interpretation?
- Are roles understood?

Systems
- Is there one clear home for this data?
- Does a data mastering strategy exist?
- Is it easy to input / submit?
- Is it easy to retrieve?
- Is it easy for users to interpret?
- Is it easy to relate one data set to another to produce the intelligence needed?
- Does the asset information system conform to standards specified by the business?
- Are there any situations in which systems can fail to process data or updates correctly?

8.2.3 The 'Do nothing' option

A perfectly valid approach to an identified data quality issue is to agree not to resolve it. This can be a valid approach if the cost/benefit of resolving a particular issue is not attractive, or this issue cannot be resolved in the required timescale. This short-medium term option will require a transparency of the reasoning for taking this course.

The 'Do Nothing' option must not be seen as a default option, but as a conscious decision taken with due governance.

The approach relies on clear understanding of the particular issues involved and an objective analysis of their impact on business credibility and legal obligations. While virtually no organisations have perfect data throughout their remit and are unlikely to be able to afford to correct all data quality problems at once, the decision to defer action must recognise and value the implications.

8.2.4 Data collection – people and technology

Current state assessment should consider whether sufficient control exists at the collection stage of the process. It should also be noted the most efficient point of checking for data quality is at the point of collection. There are both technical and people factors to be considered:

Technical
- The inherent instrument accuracy;
- Random and systematic errors (including recorder bias);
- Uncertainty in the location/ identification of the measurement; and
- Uncertainty in the recording of the measurement.

People
- Staff not providing data
- Use of wrong measurement units (e.g. metres rather than centimetres)
- Frequency of data collection and opportunities for data collection

Checklist for improvements in quality:
- Consider different instrumentation;
- Assess use of duplicate instrumentation;
- Use suitable asset identification methods (radio tags, bar codes, unique id numbers);
- Provide data validation at point of data entry;
- Consider how lists of allowable data entry values will be managed;
- Decide on automated or manual data collection;
- Consider paper or electronic recording at site;
- Identification of practical difficulties such as poor access;
- Improve training and motivation of staff;
- Consider use of a different related measurement with different inherent quality; and
- Avoid free text entry, where relevant.

The introduction of automated data collection, or of hand held data entry devices combined with electronic asset identification, can make a major improvement to asset data quality. This can not only speed up the data collection process but also facilitate automated data checks. However, as always, the cost and value of this solution must be assessed. It should also be noted that automated data collection should still be quality assessed.

8 Information Lifecycle

> *Lessons from experience:*
> *In the real world the interaction of people and technology is complex. The most successful data systems are developed as a collaboration of asset managers, frontline/field staff and supporting technology specialists.*

8.2.5 Site surveys

Site Surveys can provide an accurate method of filling gaps in datasets, although this accuracy typically comes with a significant cost. Surveys are a good tool for improving data where existing records are very poor quality or non-existent (as long as assets are not buried, submerged or otherwise inaccessible). This could be the case where new assets are adopted by an organisation for example. Whilst undertaking surveys to fill gaps in data, consideration should be given to site labelling and verifying existing data. Although this may increase the time required on site to complete a survey, if sites are geographically widespread then such an increase may be minimal when travelling time is also considered.

Surveys are typically one-off exercises, conducted by outsourced surveyors, visiting a range of sites to record information on assets, either on paper or electronically. Surveys may require access permits in hazardous areas and enabling works to facilitate the survey. Enabling works can often be more costly than the survey itself. Processes should be designed to reduce the need to revisit sites, perhaps using mobile communications to optimise the value from site visits. GIS routing techniques can minimise travel time and maximise surveyor productivity.

In order to maximise value from surveys, it is essential to scope requirements accurately and to comprehensively plan the transfer process back onto corporate systems. Digital imaging can provide an audit oversight capability to surveys if images are reliably referenced. Modern GPS enabled hand-held equipment offers great scope for remote data return for validation. Consideration of the time available for survey, the mix of team capabilities and the timeframe allowed will affect quality parameters of data collected.

Data reliability from surveys can be extremely high, provided that surveyors are appropriately experienced and trained in assessing the assets that they are being asked to survey. Such value is only realised if appropriate resources are applied to validation and upload onto corporate systems.

There is often a dilemma between a focused data collection exercise against the collection of data as part of the routine inspection/maintenance of the asset.

Whilst the collection of data on an ongoing basis is often seen as less invasive, the trade off is productivity as the maintainer must be allowed additional time to complete the job and collect / verify the data and the slower time to improve data quality. Additionally, staff will either need to be provided with specific data capture tools, or the work management system may need to be configured to allow data verification/data entry. Many companies choose increased productivity and only carry out data surveys as and when the data is required. The trade off is that you do not have the data readily available, and the cost of one off collection is invariably high in terms of survey costs and maintainer time to carry out the survey.

The challenges of formalising a one-off data verification exercise should not be under-estimated. The primary issue concerns securing the appropriately skilled resources able to conduct the exercise – and those skills must be current with recent experience in the relative technical area of expertise. If the skilled resources can be identified, the next issue is often about securing time from the individual as they are generally fully employed in an operational role. Time may be offered as a secondment to the data exercise but continuity of assignment may be challenged by 'higher priority' operational incidents. If the data exercise is significant in timescale, the Data Manager has to consider multiple secondments and the overheads in training multiple individuals on the needs of the data exercise.

> *Case Study - Water Plant Site Surveys*
> *The tiered asset inventory register for a waste water treatment plant was known for its inaccuracies. While overall plant capacities were known, limited knowledge on unit level sizes and capacities made strategic investment assessments unreliable. Surveys were scoped to collect site data, but this scope did not include the verification of the existing inventory entries against which maintenance schedule data was already annotated. Script upload of unit level survey returns thus added to inventory entries and retained old entries. Additional processes were required to gain target parameter populations.*

8.2.6 Data validation

Data validation checks should always be carried out as close as possible to the point of data entry and comprise a formal appraisal of validated parameters. Ideally users should not be able to save invalid data to a live data store; however, this is not always possible in practice. As errors are detected subsequently, so governance arrangements must determine the reporting arrangements, workflows and system facilities to enable the errors to be easily corrected retrospectively. Confidence grading of data can be a valuable tool in monitoring data validation progress and correction activity.

Automated routines for data checking can be very powerful. When data is being manually entered into an electronic system these can be, for example:
- Verification that entered data is numeric or text as appropriate;
- Pre-set bounds on valid number entries (e.g. must be in the range 0 to 100);
- Correct format e.g. formatting of dates;
- Use of drop down menus to limit choices for text items;
- Context sensitive questions (only request data relevant to the particular asset or asset type); and
- Collection of additional information to aid verification.

Paper based systems may also utilise some of these checks, particularly through intelligent use of check boxes and formatting. Whilst paper based systems have additional opportunities for human error, it must be remembered that intelligent human assessment of data can identify many data errors. Human assessors can spot data which is out of line with other measurements (either adjacent or previously recorded). The best automated checking systems will endeavour to replicate this intelligence both when entering data and when checking data later.

Often data errors are only discovered as a result of failures of subsequent transactions attempted in, or between, systems. In such cases, governance procedures must enable the data errors to be rapidly corrected and the failed transactions re-processed. The software architecture must include a transaction monitoring framework to monitor for anomalies and automatically report all instances for process audit. GIS validation is a multi-facet process requiring strict compliance to system model and processes. Design of script support within user interfaces and input data review will make a valuable contribution to GIS data reliability.

The acquisition of legacy assets at mergers and reorganisations often brings data inconsistency, which then impacts on overall metadata parameters. Audit of new data stores must use the information architecture to identify how data is to be merged into the appropriate systems, so that the likely impacts of anomalies can be identified and resolved. High-level metadata metrics provide objectivity in reporting implications of data import issues and provide a basis to monitor and intervene. Low level confidence flags can be used to track individual data records.

Checklist:
- Ensure data collection provides data to the optimum place;
- Identify checks that can be undertaken to assess data validity and accuracy;
- Evaluate technology or collection processes to ensure that they provide adequate checks; and
- Ensure that any process or technology changes are workable in the field and do not introduce other unintended quality problems.

Lessons from experience:
Some simple checks can help reduce some of the most common errors where data is entered in the wrong units and is out by a factor of 100 or 1000.

Over rigorous system validation, particularly if the validation does not keep up to date with process changes can result in users being very inventive to ensure records will be accepted by the system.

Any changes to the process which restrict data entry options need to be carefully checked. An ill-thought-out change can frustrate data collectors. Many databases contain null values or a bias towards the first item on a drop-down list because entry of the true value is too hard. Collaboration between all system users is a feature of most good practice.

8.2.7 Data infill techniques

Many business activities, such as developing long term strategies, regulatory reporting and asset valuation require complex analysis across all of an asset inventory. If data sets are incomplete, then the accuracy (and usefulness) of reports based on this data are likely to be reduced. In such cases, a variety of data infill techniques can be used to overcome gaps in data.

Data infill techniques include:
- Extrapolation - Normally a short term expedient, the extrapolation of values based on existing known values may be appropriate for particular needs to overcome gaps. The technique requires a starting point and rule base assumptions for producing a data object. Extrapolation rule bases may be necessary at different levels, in particular time frames, across regions or similar assets.
- Inference - A process by which rules are applied to determine the likely data based on other information. Inference is often applied where there are gaps in a high volume of data and related datasets can generate robust rule bases. Inference can work best for asset types where relationships are well understood. Inference is more difficult to apply where there are many possible relationships between data. This could be the case for mechanical or electrical assets, where the power rating of a pump is not just related to its flow rating.
- Synthesis - Uses recorded data to provide missing values perhaps from other recorded data. Serial numbers may contain basic size/capacity data; common abbreviations or misspellings can be recognised. There are innumerable possibilities for such data synthesis but all require intelligent knowledge of the assets and will provide differing levels of accuracy. Automated tools offer opportunities to identify and infill data.

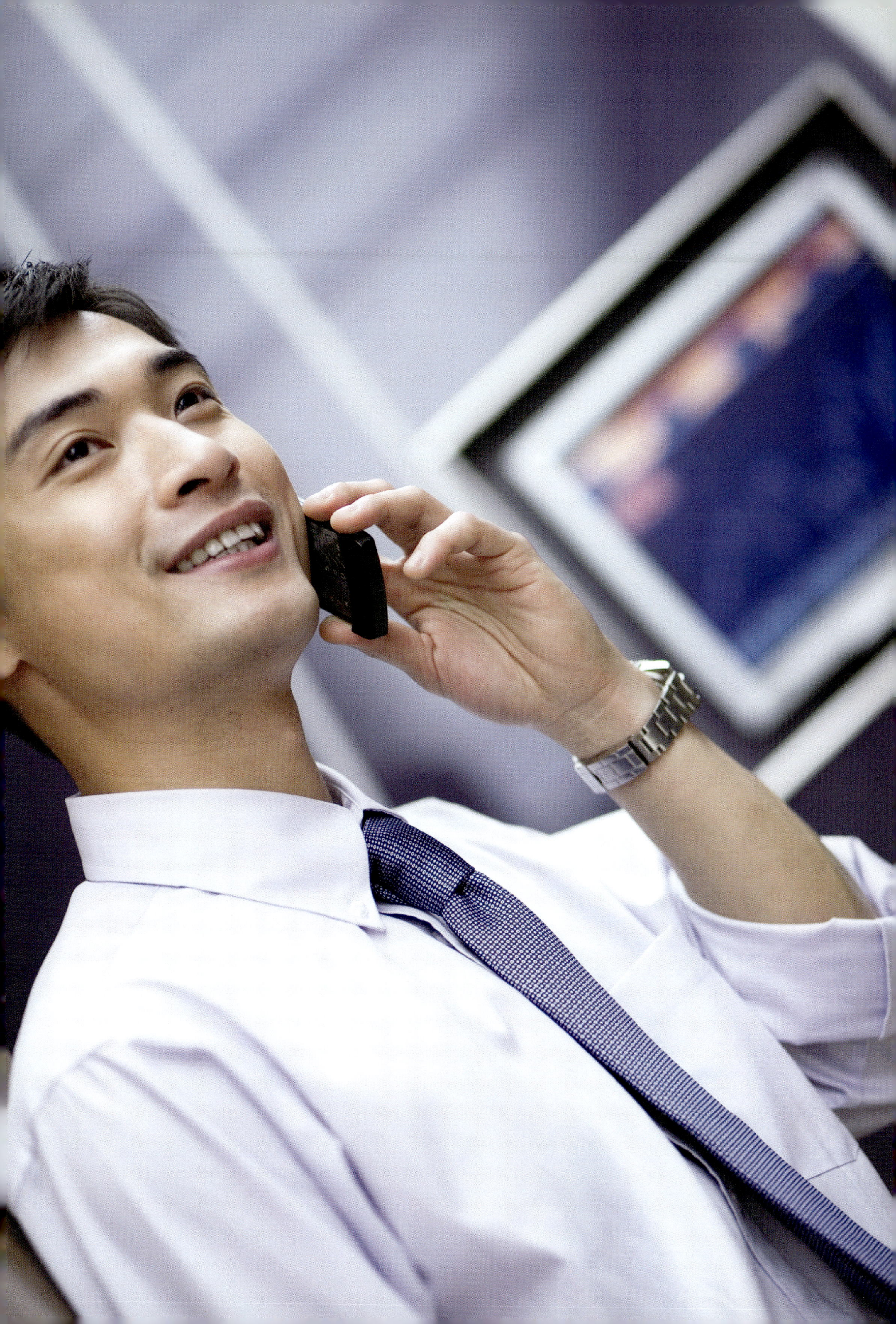

- Spatial Techniques – the use of geo-spatial proximity or connectivity allows data to be transferred from linked assets or other spatial data. The power of modern GIS and the availability of spatial data sets can provide large volumes of infill data from well designed processes. The automated nature of processes, with clear objectives and audit procedures can bring reliable improvements within a short programme.

The extent to which infill techniques may be judged appropriate requires strict governance and a transparency of assumptions. Extensive or prolonged use of extrapolation can lead to serious departures from fact and may be ignoring real information that simply requires collation. Clear identification of which values are extrapolated, inferred or synthesised is essential.

When undertaking any data infill techniques, it is important to tag which data has been infilled and how. Confidence flags depicting infill data should be used to trace the source of such values so any anomalies can be understood. It is also important that any rule base used to infill data are documented and tested to bring confidence to the process and to provide a suitable audit trail. Simple statistical analysis should support expert judgement sought across the organisation, relating historic practices, local procedures or asset characteristics.

The power of geo-spatial techniques to generate and link information has a very great potential for the improvement in population and validation of spatial data models. Like any powerful approach, good governance and strategic insight is required to obtain best value from the process. The use of confidence flags within GIS data models provides data governance at a basic level that enables the tracking of data reliability throughout acquisition, extension and maintenance of registers.

Case Study – Rising Sewers – Inferred Parameters
The valuation of pumped sewers required for regulatory reporting is based on the pipe diameter. The population of pipe parameters in 2006 was 80%, requiring 20% of valuation to be based on default sizes. Although pumped sewers would be installed in one contract, at one size at the same time, GIS operators did not annotate all pipes with attributes. In 2007 the GIS pipe elements of sewer risers were linked together by an automated process. Pipe parameters were compared and populated into empty fields by pipeline. Diameter field population rose to 96% of inventory length and raising report confidence to reliability grade A.

Case Study – Service pipe connections
A water company used inference to determine data on service pipes which typically do not appear on GIS systems by automatically creating a line from each property to the nearest water main. Where more than one main exists on a street rules have to be determined to help infer the correct main to connect to, for example trunk mains do not normally have service pipe connections, alternatively by considering the age of a main, material etc. this can help determine the correct main to connect to.

8.2.8 Option selection
Once a number of solutions have been identified, it is important to determine the most suitable option or range of options to pursue. Any decision on an initiative to change data quality needs to balance the costs of the initiative with any resulting improvements to the on-going business costs and to the risks to business objectives.

A business case is needed for a data initiative as for any other investment. This will require the input from the data users (who should have articulated the reasons for their data requirements) as well as evaluation of all the costs for any proposed change.

It is likely that the overall range of solutions agreed for implementation will be a combination of larger, project based activities and a number of smaller tactical changes.

8.2.9 Assessing the affect of data quality on business costs and risks
The benefits of any change in data quality may appear difficult to assess but a detailed discussion of data requirements with the data stakeholders should identify their value. When data was originally requested by a particular stakeholder, they should have explained its use and the consequences of poor quality data.

The benefits of improved data quality are likely to fall in the following areas:
- Reductions in risk to business performance objectives. For example, because of improved maintenance work targeting or improved operational decisions;
- Reductions in wasted work. For example, because of fewer errors in work planning or avoidance of unnecessary duplication of analysis;
- Ability to meet statutory or regulatory reporting requirements; and
- Improved accuracy of strategic risk and investment decisions.

Benefits should, wherever possible, be monetised (that is, converted into some financial measure to allow like-for-like comparison). To quantify the benefits it will be necessary to go

8 Information Lifecycle

back to the data users and to determine what costs and risks are affected by the data quality and to quantify the effect on these of a quality change. It may be necessary to seek additional data such as data on unexpected additional work, unexpected failures or aborted tasks.

Benefit examples include:

- Maintenance schedules – Inappropriate application of maintenance regimes due to a lack of understanding of asset performance. A low level of maintenance could lead to a shorter asset life and earlier replacement. Overly regular maintenance could be costly with no true benefit in terms of asset life expectations. Monetised benefits might include an estimate of the safe reduction in work due to improved data;
- Managing asset reliability – Poor quality data may lead to unexpected changes in performance of assets as changes in asset condition may not be understood and remedial work to control reliability may also be badly targeted. Quantified benefits may include estimates of changes in failure rates from improved targeting of work or simply the narrowing of uncertainty bands for predicted failures;
- Setting investment requirements/ Capital expenditure planning – Where the level of capital expenditure is based on analysis of asset performance, poor data quality could result in investment projects being promoted which are not required. At worst, this could lead to a functioning asset being needlessly replaced. At best, this could waste engineers' time by having to progress projects for assets which do not require investment. Quantified benefits would include the amount of investment at risk and an estimate of the level improvement to that risk;
- Response to alarms and operational incidents – Poor data quality could cause additional costs to organisations that provide 24/7 services where reactive work is an important element of their service. Engineers could be sent to sites to repair apparent faults which are not real (for example, triggered by faulty telemetry signals) or equipment that has failed is not reported through automatic failure detection. Customer service suffers unnecessarily whilst the problem is identified and resolved. Monetised benefits may include data on costs of attending false alarms and/or saving from having to make fewer compensation payments to customers;
- Operational efficiencies – Data quality can have a direct impact on the efficiency of operational services, for example, where maintenance crews are dispatched to carry out repairs. Data quality errors could lead to incorrect spare parts being carried, leading to revisits, and additional downtime. Monetised benefits may estimate costs of aborted journeys or reported additional work (requiring additional site visits) due to data errors;
- Capital project efficiencies – Without the correct asset data being known, the design and construction of new or replacement assets could be compromised at additional cost to the asset owner. Projects may be cancelled if the data is incorrect or not known. Alternatively, the scope of projects may not be sufficient, causing inefficiency in the design or construction process whilst omissions are dealt with. For larger projects the cost uncertainty of the project should be assessed and the effect of the data change on reducing this risk estimated; and
- Regulatory reviews – For certain regulated organisations, data quality can have a direct financial impact on their financial performance, for example, through regulatory incentive mechanisms such as the Asset Management Assessment for the UK Water Industry.

8.2.10 Putting together a case for change

Whilst considering improvements that should be made to data it is important to understand the costs, benefits and risks associated with the improvements. For more detail on the areas that should be considered, please refer to Chapter 14. Managing Change

8.3 Store data

For further consideration of the way in which data should be stored, please see Section 14 Software, which includes detail on the need for a Data Mastering Strategy (Section 14.4).

The way that information is held for an asset or asset group is a fundamental consideration that will impact on how the information lifecycle may operate. At a fundamental level, asset information may exist as an asset register, a list of physical things that are owned. Having paid for the asset, the asset register may have originated as a list in financial accounts with description, date of acquisition and historical cost being principal attributes. Extending information attributes to include location, function, condition and maintenance is likely to be generated and held within the business units that use it day-to-day. Asset information may be updated from a range of 'live' sources to fulfil asset management requirements.

The nature of the asset, whether it is static or mobile; surface or buried; of long or short life; stand alone or connected with a system, will determine the nature of its information lifecycle and the system platform(s) best suited to storing aspects of asset information. While there may be many variables across asset information, some fundamental features are common:

- A range of information attributes will need to be held to support the range of asset activities (operation, maintenance, risk management, investment, etc.) so a data model can assist in deciding how and where data is held;
- Some information attributes may be static with low levels of change (e.g. location, date created etc.) or dynamic

(status, condition etc.) using updates from 'live' systems on a regular basis, so a process model can represent flows of data;
- Specification of attributes within the data model needs to be clear so that it is applied appropriately with documented assumptions and the design of data platforms must account for the lifecycle of the various attributes held;
- Information about the data (the currency of data held, its accuracy level and reliability) often termed 'Meta-data' must determine how confidently the information can be applied;
- Modern data stores allow datasets to be interrelated across business units and may display assets spatially in context with other information to provide powerful platforms for decisions support, but are more complex to maintain even for larger enterprises; and
- A range of specialists may be involved in supporting asset information, each with their own skills and perceptions that the asset manager must command to achieve the benefit from holding and applying asset information.

8.4 Utilise data

The utilise data phase of the data lifecycle often receives a lot of attention as this is where users come in to contact with data but often this attention is misplaced. Poor requirement definition can result in data being collected and stored because it is perceived to be useful without fully considering how it will be used.

Figure 10 - Data/decision cycle

If data is being collected but is not being used to inform a decision, which in turn drives an action, then it is worth asking how the data is being used and what value is being gained from holding it.

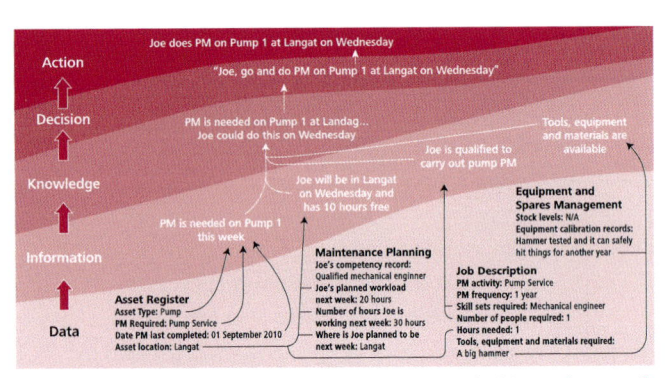

Figure 11 - Data/decision cycle example

8.4.1 Understanding Data Use

It is not a trivial question to ask how data is used; there may be answers that are not initially expected. To understand data usage it is useful to consider the following:

How is the functionality of systems affecting usage?
- Are people following the official processes and procedures or are workarounds being employed due to the software being too slow / confusing / time consuming etc.
- Is this resulting in people not using the data contained within the software at all or are they using some other method to compensate?
- Conversely, is the software flexible and accessible enough that data is being used for uses that originally were not envisaged?

How does data influence the decision making process?
- Is data being used to inform decisions?
- Is the data quality and confidence in the data communicated to the decision maker?
- Are decisions being made by users of the data in order to facilitate further actions?
- What decisions are made by customers of the data based on off-line analysis?

Who is using data and what is the mechanism used?
- Do you know all of the stakeholders? They could include other departments; balanced scorecards being used by top management; third party stakeholders such as Regulators; other software tools; decision support tools; reports etc.
- How is data distributed?
- How is information and knowledge communicated?
- What is the method of viewing / using the data?

Are security constraints restricting the use of data?
- What constraints are applied?
- Is reduced accessibility / visibility restricting data usage?
- What impact is this having on data quality?
- What is the permanency of the methods of communications and how is this restricted?
- What benefit is restricted as a result of the above?

8 Information Lifecycle

How is the data turned into knowledge and who benefits from this?
- Does system functionality facilitate this for the user?
- Is analysis required such as the application of statistical techniques or trending?
- Does other data need to be added to aid interpretation and draw conclusions?

What is the time lag between data being collected and being used to inform a decision?
- How quickly does the data element under consideration become out of date?
- How long does it take for the data contained within the system to reflect reality?
- Is real time visibility of data needed?
- What are the characteristics of the data element? For example, is it static - does not change over time (e.g. date of installation); dynamic, which does change over time but can only ever have one value that is correct (e.g. age); transactional – requires recording of events over time which will always remain correct (e.g. dates of inspection).
- Does the data have a time constraint on its usefulness? For example, is it useful to know that a wall which is red was once blue, and for how long does this remain useful?

Who interacts with data and what is their relationship to it?
- How is data created?
- How are changes to data made?
- What knowledge can be returned to the system, to update and extend information held?
- Who is responsible for making changes to data?
- Who is responsible for making sure data is correct?
- What are the consequences to each stakeholder in the data if it is not accurate enough?
- Who views the data as a "customer" rather than a data manager?
- Is the use official, documented and subject to governance or is it opportunistic?
- Are all users aware of what the data represents and is there a definition available? Is there any ambiguity that may result in misinterpretation?
- What audience is the data visible to? Is this audience internal or external?

8.4.2 Examples of data utilisation

Asset Maintenance

The use of asset information within maintenance functions has two features:
- Tactical application of data may determine the scheduled maintenance times of an asset, assist in the planning of shut-downs and reporting asset reliability statistics for performance assessment; and
- Operational response to failures may require access to specific asset information for safety, repairs and the reporting back for task scheduling.

The ability for asset information to fulfil these requirements will depend on the design of the data model, the software to support access to and transfer of data items and the training levels of personnel. Maintenance functions may be self-contained within an enterprise and require and offer little interaction with asset information, leading to a business that will fail to optimise in maintenance value and be ambushed by critical asset failure.

The availability of field devices that can hold, acquire and transmit information has opened a wide range of options to improve maintenance response and capture condition and performance information. Clearly, each function will have its own specific requirements and these are often the drivers for the development of functionality.

Inventory Valuation

The valuation of an organisations' asset inventory can be a prime driver for the retention of attribute data on assets for financial reporting frameworks. The historic cost of an asset/asset group may provide a reference point, though sustainable enterprises require a range of other bases (replacement, modern equivalent, amortised income derived) to ensure adequate provision is set aside for replacement. Depending on valuation basis, asset information on size, performance, maintenance costs, condition, asset life and installation features will influence asset valuation. Buried infrastructure can be particularly problematic as it may be so long lived that modern materials or environmental issues may change significantly from installation, with depth of assets and proximity to road types being principle variable for costing. While asset data models may be insufficiently populated with actual data, spatial techniques can associate digital terrain models, transport networks and soil cover to inform inventory valuations.

Asset Risk

Risk can be evaluated over a range of enterprise issues often using a likelihood-consequence model. Often these are driven by asset information and provide knowledge of risk, response requirements and potential costs to the enterprise.

Component reliability data can be informed by maintenance response data utilising failure incidence data to develop the probability (likelihood) of failure. Where specific reliability issues arise around a particular component, then identification of the criticality of assets affected will inform the management response. Data on equipment manufacturers and serial numbers provide the starting point for identifying assets affected and likelihood of failure. Statistical analysis can create objectivity over varying data sources to develop a robust approach to failure likelihood.

A feature of any risk qualification exercise will be the proportion of failures that can successfully be linked with particular assets, either automatically or manually. Improving the quality of failure data typically will raise the volume of data matches and thereby increase statistical confidence. Where potential for failure has a geographic aspect (ground conditions, utility strike, etc) then location of assets by spatial proximity features can inform the failure model.

Understanding the connectivity of assets within systems will determine the impact of quarantine arrangements. The consequence of failure may have wider environmental impact due to physical damage of other assets or facilities in close proximity to the asset failure. This will inform the likely cost of failures and provide the basis of mitigation to prevent failure or support response should a failure occur.

Asset Replacement

The decision to dispose or replace an asset is a key feature of its management. Leading up to this decision will draw on a range of information from the asset register and associated asset performance data to determine the remaining asset life so that replacement can be planned.

Asset life can be described financially (fully depreciated, benefit positive, high operational costs etc.), by performance (under capacity, low reliability) or by condition (failed, dangerous, etc.). While financial life is informed by cost data, performance and condition may often require a value judgement over a range of asset information, gathered from a range of sources and associated with the inventory.

Specific assets may justify a detailed study to objectively assess replacement options. Identification of which assets should be prioritised for detailed study and the number to receive such attention may require the collation of a range of information from which an initial asset life can be assessed and set in context with other asset types.

Corrosion models, for example, attempt to collate a range of parameters across an inventory to predict when critical corrosion levels may be reached. Matching projections with failures encountered in maintenance data is a key statistical test. Such models may inform an asset risk model and also offer a view of the proportions of inventory approaching end of life. Clearly, such applications of asset data require understanding of the reliability of the data used so that confidence of the indications can be objective and provide the evidence for projections applied.

Performance models may likewise draw on asset information to assess capacity and service level. Power loss and hydraulic calculations draw on asset information to represent system models that can identify critical points or pinch-points in capacity.

The dynamic nature of power and water systems and the availability of performance data from monitoring can generate large volumes of asset information held by specialist departments and often not available widely in the enterprise.

Regulatory Reporting

The provision of critical public services through monopoly structures can bring a regulatory structure where asset information influences the pricing of services or products. Concession agreements often include transfer of infrastructure at agreement termination, so condition and performance of the transferred assets will be of prime interest to government sponsors.

A key feature for regulators will be the adequacy, accuracy and reliability of the information provided by asset owners that describe inventory condition, performance, renewal and enhancement that will sustain the target consumer service under finely tuned tariffs. The potential for accusations of tariff rigging and the associated punitive fines require regulated companies to design a regulatory reporting framework and control information that they report.

The reporting regime may evolve and develop in sophistication over a series of control periods that may require asset owners to invest in a medium-term information strategy to improve the quality and coverage of their asset information. The use of metadata to describe the reliability of asset and inventory information becomes a key feature to demonstrate that information reported meets a stringent audit regime. Shareholders and other stakeholders of non-regulated industries that rely on a large asset inventory may demand similar stringency for their enterprises.

The demand for asset information by regulatory reporting regimes may be more stringent that normally required by operations in service units. To avoid two or more versions of information being in circulation within the enterprise, a regulatory standard for information management may need to be adopted to retain efficiency. The potential costs should be balanced by the benefits in improved accuracy of information for decision support and improvement in asset knowledge.

8.4.3 Utilisation Measurement
Benefits
Once it is understood how data is used within an organisation, it can be beneficial to try to measure usage. Such benefits include:
- Increasing understanding of the value derived from the data;
- Identifying critical datasets and assessing the associated risk;
- Identifying data that is not being used;
- Identifying areas of waste such as user licences that are not being used;

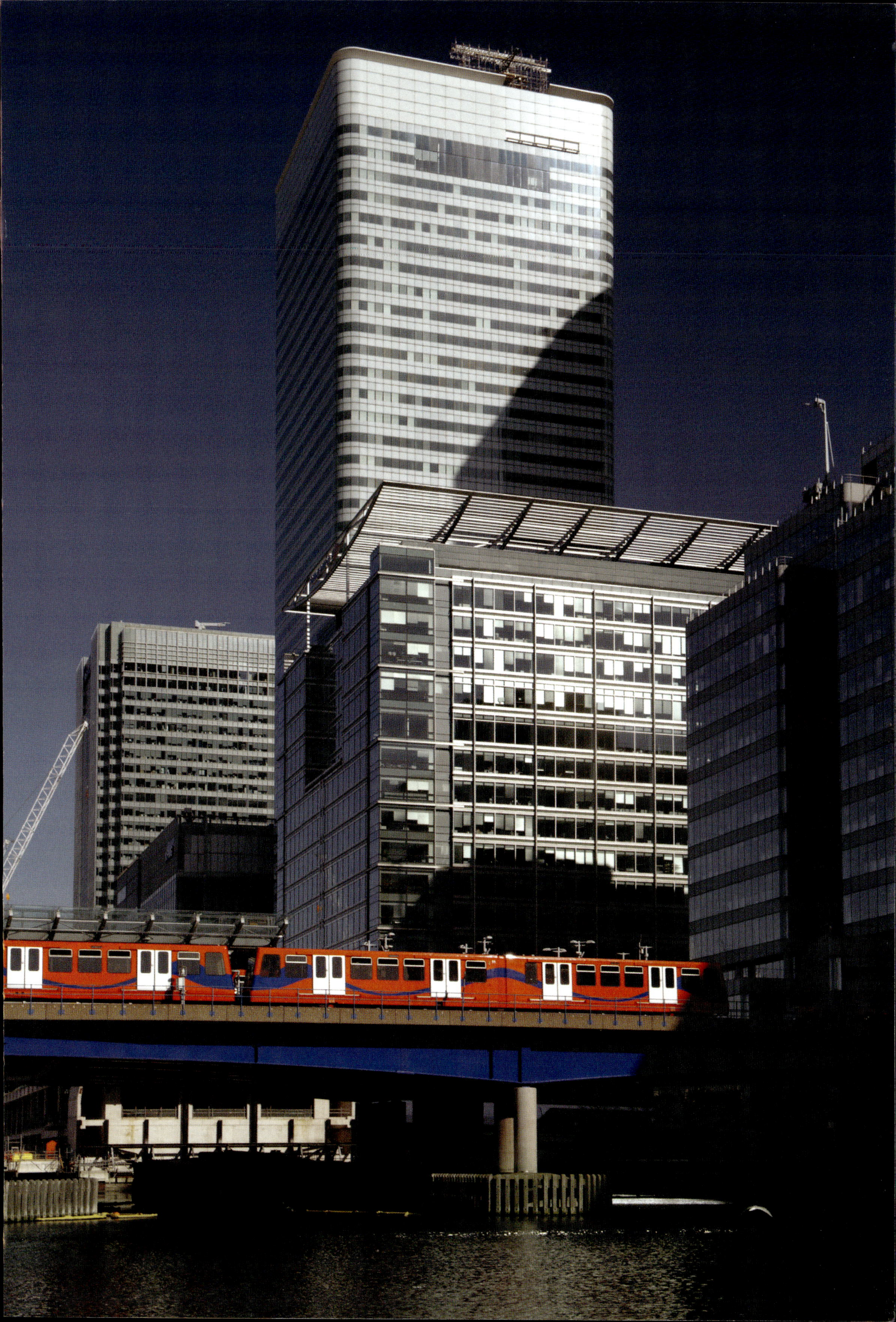

- Helping to ensure that data is being used as intended and interpreted correctly; and
- Ensuring people make use of the data being held within the organisation.

Techniques

There are numerous techniques that can be used to measure data usage, the selection of which will be heavily dependent upon the dataset being considered and the method of utilisation:

- Traffic;
- Data uploaded / downloaded;
- Metadata queries identifying access times, tracking changes etc.;
- User logins;
- Time spent logged in to systems;
- Number of reports run;
- Number of database queries;
- Number of times documents have been checked out;
- Page views;
- Click through rate; and
- Number of links / references.

Consideration should be given as to which method of measuring usage will give the best indication of usage, how valuable this interaction was for the organisation and the costs associated with measuring the data.

8.4.4 Improving Usage

Increasing the use of data held by an organisation not only results in more value being derived from the data but also helps to improve data quality. A data user will have an incentive to ensure data is correct and therefore will pay more attention to ensure errors are corrected and it is managed properly. If data usage is understood it will be possible to identify reasons why data is not used. These barriers to use can then either be addressed to ensure that the intended benefits are realised or consideration can be given to archiving the dataset.

A common barrier to data use that also results in duplication of data is people not realising that data exists within the organisation and not being able to access it. Therefore increasing visibility of the data that an organisation holds and making this available to the widest possible audience can be one of the biggest and easiest methods of improving usage, often in areas that would not otherwise have been considered.

8.5 Acquire new data

Asset information most often is acquired on asset commissioning, but also can be as a result of surveys or be derived by transfer from other business units as a result of organisational change.

All acquisition processes require a clear understanding of data specifications, the process for acquiring data and where the data will be stored. Issues of accuracy and reliability are relevant for each data item, with coverage and population being relevant for the asset inventory as a whole.

Asset creation/commissioning

To provide good quality asset information, it is important that the process of acquiring asset information from asset creation/commissioning is one that runs alongside the physical construction/commissioning processes. Historically, organisations have tended to gather data as a single activity prior to asset handover, which typically is ineffective, inefficient and results in poor quality data. Key considerations are:

- Ensuring that information provision is considered as important as the provision of the assets themselves;
- Information provision is a managed process throughout the life of the project;
- Data is acquired from design and procurement information wherever possible, in order to minimise site data gathering activities;
- Data acquisition should be by purpose built data forms/ applications wherever possible;
- There should be easy access/ visibility of the Asset Data Dictionary (see Section 6.4);
- Project handover/ sign off should not be possible if asset information has not been provided; and
- Asset information governance should be closely focused on monitoring provision from asset creation activities.

Surveys

Physical surveys of assets provide a clear information stream to populate the data model of the asset inventory with observed features. See Section 7.2.4 for more information.

Transfers

Asset information may be transferred from other business units or inherited from legacy organisations / third parties. In such cases the data should not just be considered as a 'gift'; the data should be processed as part of a formal data migration project, see Section 13.5.4 for more information.

Spatial Context

The availability of Geographic Information Systems (GIS) at corporate, tactical and operational level allows an asset register to be expressed visually, placing assets in a geographic context with other assets, the demand for services and other environmental information. Spatially distributed assets (rail, energy, telecoms and water networks) have system connectivity features and interact with a range of map features. Spatial visualisation also supports the acquisition screening, so that anomalies, duplication or missing features can be readily identified for resolution. While not replacing an asset register, a spatial facility offers additional functionality for meeting a

8 Information Lifecycle

range of data, regulatory, risk and investment tasks, not least the facility for associating information about asset performance (e.g. failures) with specific assets. This provides an additional route for the acquisition of asset information to create a wider asset knowledge.

There are many different ways of visually representing relationships between assets. Thought should be given to the purpose of a diagram in order to communicate the relevant information. For example, a topological representation such as the London Tube map communicates to the viewer relevant information such as the number of stops between destinations and where to change lines so that the most efficient route can be planned. However this map alone does not allow a properly informed decision to be made as you cannot tell from the Tube map how far it would be to walk to your destination above ground from an alternative station; a geographical map would be needed to determine this. Conversely, it would be very difficult to quickly plan a journey on the Tube using a geographical map as the relevant information is not clearly visible. Displaying a map of such a large network clearly at an appropriate scale would not only be problematic but further bury the relevant information within other, non-relevant information.

8.6 Archive data

Data archiving forms an important part of the asset information lifecycle and is often not considered in sufficient detail. There are many different types of data sets that hold asset information and each will have their own unique requirements. Questions that should be considered when deciding upon archival requirements include:
- What should be archived and why?
- For how long should records be retained?
- What are the obligations, legal or otherwise, for retaining records? These could include regulatory, statutory and legal obligations as well as internal business requirements;
- Where there are obligations related to data retention, who is responsible for ensuring these obligations are met?
- What are the risks associated with not being able to retrieve records?
- How quickly archived records need to be retrieved?
- How will archived records be accessed?
- Who should be able to access archived records?
- What lockdown controls are needed to ensure archived records have not been changed?
- What metadata should accompany the record to reconstruct timelines and prove validity in the event of a dispute?
- What long term storage media should be used? and
- What value is there in retaining records and how does this value change over time?

When these questions have been answered it should be possible to formulate a data archive strategy. Such a strategy should seek to balance the benefits from retaining data with the costs involved in the archival mechanism.

As well as developing a thorough archive strategy, it is important to consider the business change implications of implementing the strategy. It is essential that employees are educated to the reasons why records are being archived, how they can access this data and for how long. It may also be necessary to challenge perceptions of value, especially if this is likely to diminish over time. If these messages are not communicated properly then it is likely that some of the primary reasons for archiving the data set will not be achieved. For example, data proliferation may occur as a power user, fearing the loss of a dataset, downloads everything they have access to and stores it offline in their own, personal database. Not only is this likely to take up more storage space than the original data set but this situation may be replicated around the organisation and result in out of date, uncontrolled data sets being used.

8.7 Delete data

The final, permanent part of data disposition, deleting data, is often considered even less than archiving data. In many cases there is a natural resistance to deleting data as people like the comforting thought of having the ability to refer to archived data if they ever need to. In addition, this behaviour is often not discouraged as the rate at which storage capacities have increased and associated costs have decreased have not provided enough of an incentive to counter the perceived benefit of holding archived data. Interestingly, the costs associated with data storage are often included in the budgets of parties who are not responsible for making the decision whether the data should be retained or deleted, which also does nothing to discourage data "squirrelling".

A data archive strategy should specify a retention period, after which the data should be deleted. There should be a process that is followed in order to delete data which includes obtaining authorisation from the relevant parties, including those with responsibility for ensuring obligations are met. It may be necessary to consider security requirements surrounding the deletion of data and questioning what disposal method is appropriate to employ. This will be dependent upon the sensitivity of the data being deleted. If certain disposal methods are mandated to ensure permanent deletion, this should form part of a wider information management policy that also governs the disposal of hardware and be included in the data deletion process.

8.8 Sustain data quality

If significant time and effort has been expended acquiring good quality asset data, particularly if physical access to the asset concerned is difficult, then it is vitally important that the quality of this data is sustained and possibly improved. If data were to be allowed to degrade, then in most circumstances it is likely that the cost of sustaining good quality data would be significantly less than the impact of poor quality data on decision making, regulatory reports etc.

If poor data management practices exist, for example, temporary staff entering data without suitable quality checking and oversight, then data errors may arise which may not become evident for some time, by which point it is not possible to recover good data. As such, poor (or missing) data updates can create a sustained and insidious decline in the quality of data and the value that can be gained from it. Although data backup up strategies can recover data in the case of server failure, for example, these will typically only be able to recover data to a point in the recent past (24 hours up to possibly 3 months) however, this will not enable an organisation to correct data errors that have arisen over a long period of time.

If data controls are weak, and staff recognise this weakness, then poor data practices can spread rapidly across a whole organisation speeding up the rate of data quality degradation. Additionally, any investment previously undertaken to improve data could be viewed as abortive expenditure if data errors are allowed to arise.

Sustaining good data typically involves many of the activities listed in this SSG:
- Clear data requirements
- Monitoring of processes and data
- Improvements to behaviours, processes and governance
- Audit to determine what really is happening

8.9 Key Messages

i. Information has a lifecycle, similar to physical assets, that requires different techniques to be utilised at different points in its lifecycle

8.9.1 Assess data

ii. Organisations should assess the availability of information to different systems, processes and users

iii. Assessing the current state of information quality will require the utilisation of one, or more assessment techniques

iv. Data quality requirements should be agreed, defined and published prior to assessing data quality

v. Data quality is typically assessed using the metrics of accuracy, completeness, validity, consistency, uniqueness and timeliness.

The assessment of quality will differ depending on business usage and may require assessment of a number of different metrics
- Accuracy – data correctly represents the asset it relates to
- Completeness – all assets and attributes are populated
- Validity – data is stored in the correct format
- Consistency – an asset has the same identifier across data stores
- Uniqueness – each asset is recorded once, and once only
- Timeliness – minimal time delay between an asset changing and the related data change

vi. Accuracy of data is arguably the most important data quality attribute and can only be assessed by checking the data against the physical asset it represents, however, the cost and ease of assessing accuracy should be considered before undertaking accuracy assessment activities

vii. Organisations should assess the completeness of their asset inventory and the completeness of asset attributes

viii. Profiling tools can be used to assess the validity, consistency and uniqueness of data

ix. Data quality assessments should be communicated clearly and unambiguously to stakeholders

x. The activity of assigning confidence grades to data will improve understanding of the reliability of data

xi. Most of the effort required to make an assessment of what is already in place should concern evaluating the quality of what is available, rather than whether such information exists.

xii. In order to measure the quality of asset information, reusable and accurate assessment mechanisms should be put in place

xiii. The level of assessment scrutiny should be appropriate to the information's criticality in terms of its urgency and importance to Asset Management

xiv. In addition, the monitoring activity will look to identify the root-causes of information quality and gaps given that the Information Governance framework details what should be occurring by whom, when and how

8.9.2 Improve data

xv. The benefits of data improvement activities are typically diffuse and fragmented, therefore time needs to be spent to identify and quantify all relevant benefits

xvi. The costs of data improvement should be estimated as accurately as possible and should include software costs, costs of improvement, training costs and ongoing costs to sustain quality

xvii. A clear business case expressing the costs and benefits of improving data quality should be submitted for approval including assessment of the relevant payback periods and rate

xviii. A data quality framework can be a valuable mechanism of managing the improvement of information quality

xix. Root cause analysis should be used to try and determine the underlying root cause of data issues in order to ensure that sustainable improvements can be delivered

8 Information Lifecycle

xx. A valid option to assess is to "Do Nothing" as not every issue will be worth resolving
xxi. Site surveys can be used to gather missing data or to validate data, but are expensive activities, so should be planned and executed carefully
xxii. Improvements to data validation practices help to ensure that invalid data is not entered into systems
xxiii. Various data infill techniques can be used to fill gaps in data without requiring physical asset surveys, but it is important that data gathered in this way is clearly marked as such
xxiv. Changes to technologies and systems used can improved the management and manipulation of asset data
xxv. Changes to organisations can help ensure that data update activities are appropriately managed
xxvi. Changing the culture and behaviours towards data is a valuable activity, but can require ongoing and sustained effort to achieve

8.9.3 Store data
xxvii. A planned approach to data storage should be developed and implemented to reduce the likelihood of duplication or inconsistencies
xxviii. Data stores should be clearly defined with meta-data recorded for the information being stored

8.9.4 Utilise data
xxix. Good quality data and information can help deliver better quality business decisions which should lead to better outcomes and probably good quality output data
xxx. It is important to understand how data supports decision making processes, the quality of the data being used for a decision and the sensitivity of analysis to degraded data quality
xxxi. It is important understand the data being used for a decision, the timeliness of this data and how other processes may also use this data

8.9.5 Acquire new data
xxxii. The majority of new asset data arises from asset creation, refurbishment and overhaul activities
xxxiii. It is important that clearly defined, proactive processes are set up to acquire data
xxxiv. Monitoring of data provision activities should be one area of focus for data governance activities

8.9.6 Archiving and deletion of data
xxxv. Clear policies should be agreed for how long different items of data should be stored
xxxvi. Archiving could be to on-line storage or off-line storage with different requirements for accessing data
xxxvii. Once data has been deemed to be of no further use to an organisation it should be physically deleted following agreed processes

8.9.7 Sustain data quality
xxxviii. The cost of sustaining information once it is created is typically less than the risk and potential losses of failure to maintain information, or the retrospective cost of improving information
xxxix. Loss of value to information can be rapid once arrangements to sustain it break down, so that it is important to set up robust and continuous arrangements.
xl. The information and its risks need to be identified in order to target effort and plan for it to be sustained where it is most required.
xli. Information is inextricably linked to the people and processes that create it and systems in which it is held. A wide range of roles from those involved at the coal-face to those managing IT Services and Suppliers are required to sustain information effectively.
xlii. It is key to allocate ownership of information to asset managers in the business who work together to ensure all other arrangements are put in place to sustain its integrity. Due to the "locked-in" nature of information to every part of the business, management arrangements that touch every part of the business are required to minimise the degradation of vital information through neglect or dysfunctional behaviours.

9 Monitoring

Monitoring of asset information is analogous to condition assessment of physical assets, and equally challenging to report quantitatively. The development of the monitoring process however enables the key areas of information and key data to be monitored. It provides visibility to weaknesses in the data and software that would otherwise be hidden and unaddressed until the risk they represent is realised when used in anger.

Section 8.1 detailed techniques to Assess Data, primarily as a one-off activity, however, the ongoing monitoring of data quality activities is a key input to the Data Governance function (see Section 5).

9.1 Data quality measures

The measures used for ongoing data quality monitoring are likely to include a subset of those used as part of a formal data quality assessment (see Section 8.1.2) but are also likely to include measures relating to data provision processes, data utilisation etc.

Due to differing circumstances and organisational requirements there is no single set of measures that can readily be applied to all organisations. Typical measures are likely to be based around the key data quality dimensions of:
- Accuracy – Based upon physical asset inspections; and (data) non-conformance reports
- Completeness – Perhaps using automated tools and processes; and
- Validity – Use of data profiling and other system based checks of validity.

Note that, whilst other data quality dimensions may be relevant to an organisation, they are likely to be less suitable as a basis for ongoing performance measures.

Organisation specific measures of overall data quality and progress in improving data should be developed based upon the particular needs of an organisation. There should be a recognition that many of the measures that are developed will continue to be refined and enhanced over time, as greater experience in their effectiveness is obtained. This means that the tools used for monitoring and measurement will need to be flexible enough to easily allow ongoing changes to measures, as they are developed and refined.

Monitoring can be considered as an activity additional to day-to-day asset management activities. Asset management staff are using data all day long and may be finding quality issues.

Intelligent data quality monitoring will draw on this knowledge by using data non-conformance reports and other user feedback mechanisms to provide an active monitoring capability.

Prior to making any changes to measures, it is important that the measures are tested, reviewed and approved by the Data Governance function. Monitoring of changes to performance as measures are changed should be undertaken to ensure that the measures are working as intended.

9.2 Targets

An important point to consider when undertaking a data quality improvement initiative is the level of data quality required. Relevant targets should be established in order to quickly identify whether performance is acceptable and may require different targets for different business areas. By setting different thresholds, targets could be illustrated by means of colour coded 'traffic lights', in which case different target levels will be required for different colour indications.

In certain circumstances, there may be a limit to the extent of data quality that is possible to achieve. This could be due to inaccessibility of assets for example. Such limitations should be considered when setting data quality improvement targets which can then be set to be realistic and achievable.

Over time, as performance and quality improves, there will be a need to adjust targets in order to continue to drive improvement and to incentivise staff.

9 Monitoring

9.3 Compliance

'Compliance' is the term for the task of documenting that activities and outputs conform to set standards. It is important that organisations agree and state which information standards that they need to comply with (see Section 7)

Relevant standards can include:
- Legal standards for recording data and information;
- Regulatory standards for information;
- Interoperability standards to allow systems and organisations to share information;
- Technical standards to allow systems to function; and
- Internal policy and quality standards.

Once the list of standards that need to be complied with has been agreed, this list should be published to ensure that relevant staff are aware of the standards to be followed and the compliance procedures for conformity. Organisations should assess their level of compliance with these agreed standards in order to confirm levels of compliance and in order to identify any new data gathering or improvement activities that may be required. It is more cost effective to ensure that data is correct at the point of collection and data entry, than to undertake wide scale data improvement activities.

9.4 Presentation of Measures

Presentation of measures is an important activity to allow key stakeholders to quickly assess the current position and progress of data quality activities.

The expected presentation could convert the quality measures into scores which could be aggregated into logical groupings that have meaning for an organisation, this could be in the form of a 'scorecard' to include a number of tiers of more detailed measure information representing the lines and areas of responsibility. Presentation would include targets if they have been previously agreed and an indicator of Red Amber Green (RAG) status. To aid the reader, the presentation may also include trending indicators (i.e. arrows to indicate improvement from last period, degradation or same as last period) or graphs of current and historic measures.

Where deviations from planned progress or performance are detected, then these reports should allow management functions to detect these deviations in a timely manner and then to agree firstly, whether any remedial action is required, then secondly to agree the remedial actions, their timescales and the key responsible staff. Ongoing monitoring of these remedial actions should be included in subsequent performance reporting to ensure that agreed timescales and objectives are achieved.

The meaning of measures and the level of 'acceptable' quality appropriate for the operation/process should be explained. It needs to be clear which aspects of data must be 'perfect' (due to consequences of poor data) and where (for example) a lower level of quality is 'sufficient' for the operational use of data.

Frequency of reporting will be driven by the need to balance the need for regular monitoring, the costs and ease of creating reports and the time for changes to take effect.

9.5 Key Messages

i. Ongoing monitoring of actual data quality and the performance of related processes is an essential activity to provide visibility of emerging issues and the rate of improvement

ii. Organisations should agree on the key measures that they wish to monitor

iii. Targets should be set to encourage improvements in performance

iv. Targets should be SMART, i.e. Specific, Measurable, Achievable, Realistic and Timely

v. Once established monitoring should be undertaken on a periodic basis and the results fed back to asset information users

vi. The presentation of measures and different visualisation techniques should be chosen carefully to ensure that they provide required measures in suitable formats

10 Audit and Assurance

Audit and assurance enables you to review your data, software and processes objectively against standards and requirements. Using this approach you can evaluate if your data fits within expected ranges. An organisation with a mature and effective asset information management system will gain greater confidence in their data.

10.1 Audit

Data management activities should deliver good and improving data, and the ongoing governance and monitoring of data should ensure that changes to quality and new issues are identified and addressed. However, some new changes may not be identified, or there may be deviations from intended approaches. In order to ensure that such issues are identified, quantified and subsequently resolved, it is essential that a suitable audit programme is instigated.

Whilst there are many references sources and standards for auditing activities, the following key factors should be considered from an asset information perspective:
- Sample size to include enough assets and attributes to ensure that there is statistical significance in the results; if discrepancies are identified, an even larger sample may be viewed to determine the magnitude of the discrepancies;
- The selection of assets to audit should be randomised (to ensure that geographic coverage is unbiased) and stratified (to ensure that the proportion of, say, asset classes audited is comparable to the proportion of total assets);
- Auditors should be familiar with the technology of the assets being audited, but not with the specific assets themselves; and
- Data quality problems are often the symptoms of other business issues. Auditors should attempt to determine the true root cause(s) of data quality issues.

It is suggested that the following specific areas should be audited:
- Data audit – Formal audit of the quality of data to current asset information standards;
- Process audit – Audit of business processes that use/ create/ update / interact with asset data. Audits should also ensure that process roles and responsibilities are understood and carried out correctly and that all relevant standards and legislation are complied with;
- Culture audit – Audit of organisational culture to assess whether a supportive or a blame culture exists, to identify cultural barriers and competing pressures which adversely affect asset data quality;
- Data quality audits as part of other business audits – For all other relevant business audits their scope should be reviewed to identify areas where data management and data quality can be assessed as part of these audits; and
- Annual data quality audit – A regular review of data quality should allow an organisation to track changes in data quality over time. This will inform the Data Governance function and allow it to check that the rate of improvement meets organisational objectives.

Where relevant, the presentation of data geographically can be particularly effective for spatial data, and performance data can also be presented on a time line to quick visual assessment that can again pick up many simple data entry errors. Both methods have capacity for automation and subsequent real-time data entry checks.

It should be noted that in certain industries, such regulated utilities, there may be a legal or regulatory requirement for external auditors to assess the organisation. Such audits may cover a variety of processes and activities, but are also likely to involve the provision of data extracts and analysis. In order for such audits to proceed effectively it is important to ensure that there is suitable support to organise meetings, data provision and to progress any failures to provide information. Where data needs to be supplied, it is prudent to include an assessment of the quality of the data provided and the mitigating actions that may be in place when using such data.

> *Example:*
> *An Asset Process Improvement Initiative at a Manufacturer in a regulated industry failed a regulatory audit due to the lack of standard naming conventions and the inability to produce documented evidence of compliance quickly, which resulted in an 18 month shutdown.*

10 Audit and Assurance

10.2 Assurance

Assurance can be defined as the processes that assess the level of compliance with standards and the effectiveness and efficiency of business policies and activities. Assurance activities are an important part of the work an organisation undertakes to monitor compliance with standards and to improve process resilience.

It is important that competent staff are identified who are able to undertake assurance activities and whose reporting lines minimise the risk of any conflicts of interest. Assurance activities can include audit, ongoing measurement of data accuracy and establishing reports to measure the effectiveness of data update processes. The outputs of assurance activities should be one of the inputs into overall governance processes.

10.3 Key Messages

i. Suitable audits of both data quality and data related processes will help identify any deviations from planned approaches
ii. Audit measures against standards to gauge objective progress, benchmark against peers to motivate change.
iii. A specific responsibility to provide ongoing assurance of asset information processes will have been established to provide assurance that company standards are being followed and that appropriate controls are in place to maintain compliance

11 Benchmarking

The effective use of benchmarking can be particularly useful when justifying your data to regulatory bodies and can raise expectations of what performance is achievable through seeing what other companies have delivered.

11.1 Why benchmark?

Within regulated industries there is a constant requirement for improvements in performance and for the ability to demonstrate that improvement. Likewise, other organisations also have drivers to improve performance. Without a suitable measurement process, such as benchmarking, it can be extremely difficult to identify an organisation's current performance level against its peers.

A suitable benchmarking process is essential for any organisation reliant on asset information management. The benchmarking process should provide a reference for the current performance of the organisation which can then be compared with any previous assessments and with current good practice approaches. This will allow organisations to gain a view on where improvement is required, the level of improvement required and the rate at which performance is changing based on any previous assessment.

11.2 Benchmarking and change

Many benchmarking initiatives fail to deliver performance improvements as they identify difference but have no mechanism for effecting change. To deliver business benefit, it is important that both a team and clear objectives are established so that any differences are followed through and a change programme initiated to overcome the inevitable difficulties of a "not invented here" syndrome.

Once benchmarking has identified areas of potential improvement, a second stage is required to understand the factors underlying the better performance in one Organisation rather than another, and consideration given as to how that experience can be translated and indeed whether that experience can or should be translated.

The change initiative team (which doesn't need to be full-time) can also make use of qualitative assessments. For example, two asset classification systems may not be directly comparable; however the qualitative or mapping comparison of the structure of linear asset records by the benchmarking team may provide transferrable lessons and be incorporated into the change programme. In this way, benchmarking is not simply an exercise in demonstrating bare "numbers" but a living mechanism for change.

The benchmarking and assessment methodologies discussed in this section should also be considered in conjunction with the methodology described in earlier sections for identifying requirements, assessing options, and delivering change – and be an integral element of the asset management improvement programme. For more details on Managing Change, see Section 15.

11.3 What is 'Good Practice'?

Good practice is not a fixed destination, but a relative position based upon where performance of the leaders in an industry currently are and the perception of what should be achievable within that industry. Whilst theoretically one organisation could demonstrate Good Practice performance in all areas, it is more likely that the good practice performance benchmark is a combination of scores from a number of organisations.

Good Practice levels should also be determined based upon comparators from other industries, nations or sectors. These broader measures should be included to ensure that, where an industry may have low levels of performance from all participants, that the notional best practice levels are not artificially low.

Due to changes in technology, management techniques, expectations and asset management processes the definitions of best practice will change over time. These changes in Good Practice levels are likely to undergo steady, incremental changes; however, it is not inconceivable that a 'step change' in Good Practice could be achieved.

11.4 Assessment guidelines and maturity modelling

There are a number of factors which should be considered when planning and implementing a benchmarking assessment process. There are other reference guides available that can describe the process in more detail, but some of the key recommendations include:
- The process should be supported by documentary evidence or direct observation;

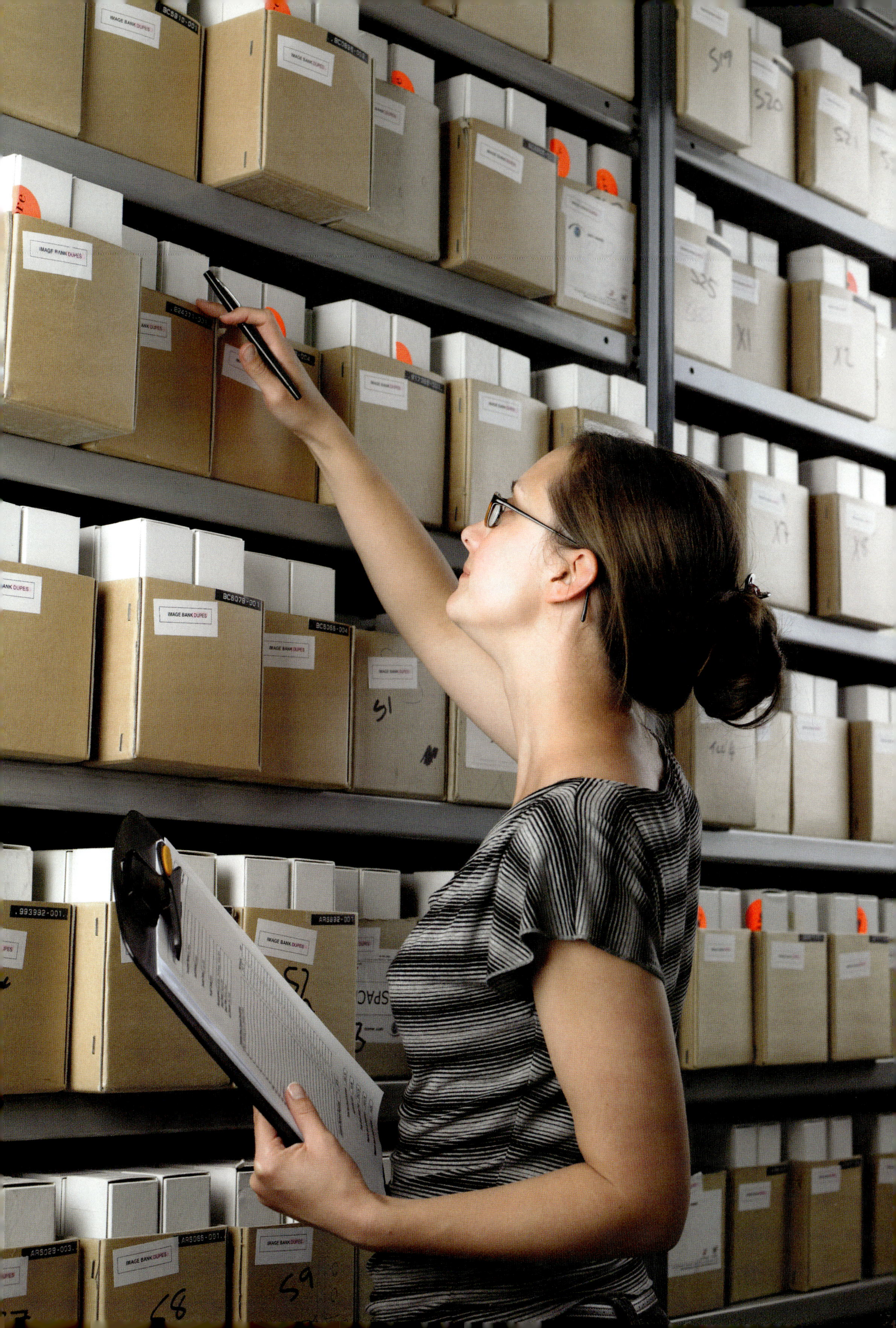
IMAGE BANK DUPES

- Assessment should cover a cross section of individuals at different levels in an organisation;
- Assessment questions should be open and free of bias;
- The process should allow auditing, if required;
- Outputs should facilitate gap analysis and development of improvement actions; and
- The process should allow the tracking of improvements through future assessments.

As outlined above, many of the factors related to Asset Information Management cannot be measured easily and repeatably due to the nature of the subject matter. A well designed process will ensure a level of consistency and repeatability between organisations and industries and most importantly should ensure repeatability between different assessors. Subsequent analysis and "lesson learning" activity can also include qualitative comparisons that the change improvement teams can translate back into their own cultures.

A number of proprietary maturity modelling techniques exist that allow repeatable and consistent assessments. Typically these maturity models will utilise an assessment criteria which has a number of related parameters and statements related to each maturity level. The assessor then chooses the level that best represents the current maturity of an organisation and records any supporting evidence and comments. Results are then combined to reflect an overall maturity level.

11.5 Alignment with the IAM Self-Assessment Methodology Plus (SAM+)

The Self-Assessment Methodology Plus (SAM+) has been developed to enable organisations to measure their conformance with PAS 55, ISO 55001 or the AM Landscape, and is designed to promote good asset management practice, embodying the quality principle of continuous improvement.

SAM+ utilises Maturity Scales to assess alignment with the chosen framework and incorporates 121 questions for PAS 55 or 126 Master and sub-questions for ISO 55001 or 320 L3 Criteria for measuring capability and maturity against the 39 Subjects of the AM Landscape. These questions/criteria are posed and maturity is assessed against five pre-determined maturity levels. The maturity scale for ISO 55001 is shown below.

SAM + questions/criteria are deliberately designed to assess conformance only and therefore limits the range of maturity levels that can be applied as L0 to L3. NB: L4 (optimising) and L5 (excellent) have been combined and are referred to simply as 'Beyond'.

The maturity scale also adopts the following principles:
- When conducting assessments evidence builds from left to right on the maturity scale;
- As indicated by the colour transitions, the boundaries of the maturity scale are not hard values;
- Conformance lies within Maturity L3 however this is not a 'pass' or 'fail' numerical value but lies within the dark blue zone.

SAM+ contains multiple asset information and data-related questions/criteria; however the purpose of an assessment is to measure the 'overall' conformance of the AM System against the chosen framework. Therefore the asset information and data-related questions/criteria should be viewed in their 'overall' context of asset information within the wider AM System, not just considered in isolation.

Innocent	Aware	Developing	Competent	Optimising	Excellent
Maturity Level 0	Maturity Level 1	Maturity Level 2	Maturity Level 3	Beyond	Beyond
The organisation has not recognised the need for this requirement and/or there is no evidence of commitment to put it in place.	The organisation has identified the need for this requirement, and there is evidence of intent to progress it.	The organisation has identified the means of systematicaly and consistently achieving the requirements, and can demonstrate that these are being progressed with credible and resourced plans in place.	The organisation can demonstrate that it systematically and consistently achieves relevant requirements set out in ISO 55001.	The organisation can demonstrate that it is systematically and consistently optimising its asset management practice, in line with the organisation's objectives and operating context.	The organisation can demonstrate that it employs the leading practices, and achieves maximum value from the management of its assets, in line with the organisation's objectives and operating context.

Figure 12 SAM Maturity Scale for ISO 55001 and the AM Landscape

11 Benchmarking

The output of assessments can be displayed in either Radar Chart or Bar Chart formats. Below is an illustration of an ISO 55001 assessment.

Benchmarking of asset information systems and processes, and implementation of subsequent improvement plans, will assist in moving towards overall ISO 55001 conformance. Assessments can be carried out internally, but runs the risk of outputs being biased based on limited perspectives and local agendas. The support of an external capability can help to keep the assessment objective; it can be a source of comparative data which may be even more comparable if carried out by the same assessor.

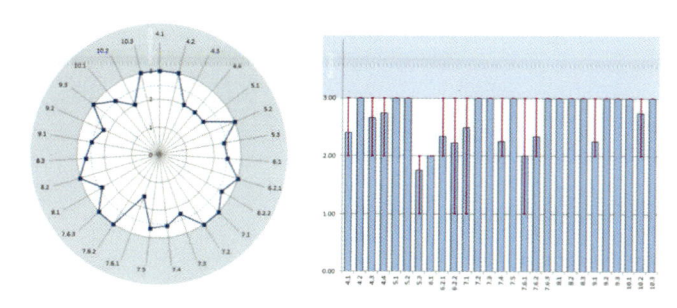

Figure 13 – IAM SAM+ outputs

Main Topic	Sub-Topics
Asset Information Strategy Requirements	• Asset Information Requirements • Asset Information Strategy • Technology Strategy • Asset Information Plan • Governance arrangements • Business Continuity
Core software	• Asset Register • Plan and Document Management System • Maintenance Management System • Failure & Performance System • Finance System • Asset Condition System • Programme Management System • Operational Work Management System • Geographical Information System (GIS) • Management Information System
Decision support tools	• Asset Deterioration Models • Whole Life Cost models • Asset availability model • Maintenance Optimisation • Demand Forecasting • Capital Planning System
Support software	• Spares & Materials • Asset Utilisation • Risk Database • Authorisation management • SCADA and telemetry
Data Management Standards & Procedures	• Data Quality Standards • Data Population Plan • Data Dictionary/Data Quality Register
Population and Quality of Data in Core Systems	• Asset Register • Plans, Drawings and Documents • Location and/or Connectivity Data • Maintenance Management Data • Failure and Performance Data • Finance Data • Asset Condition Data • Programme Management Data • Operational Work • Management Data
Population and Quality of Data in Support Systems	• Spares and Materials Data • Asset Utilisation Data • Risk Data • Authorisation Data • SCADA and Telemetry Data • Condition Monitoring Data
Management Information	• Management Information Skills • Analysis Tools • Management Information Reports

Figure 14 – Asset Information Good Practice assessment areas

11.6 Asset Information Subjects for assessment

Figure 14 below illustrates example areas that could be assessed as part of a benchmarking process for asset information.
It should also be noted that within most industries and organisations, there will be a steady change in the systems and technologies used to deliver the outputs of the organisation.
A good benchmarking process should focus on the utilisation of technology and not the technology itself. This is particularly relevant when, for example, comparing an organisation that is using old technology highly effectively with an organisation that has implemented cutting edge technology less effectively.

11.7 Key Messages

i. Benchmarking of other organisations can provide an awareness of areas of strength and weakness
ii. The IAM has published a methodology for assessment against BSI PAS 55 which is a helpful reference
iii. Specific Asset Information assessment criteria need to be developed, and example criteria are given in Figure 11
iv. A structured process is required to ensure validity of outputs and repeatability of process;
v. Qualitative and comparative assessments integrated into a change programme can be motivational and address the underlying culture required to make change happen
vi. The assessment process can provide an overall framework for asset information improvement, and can be used as a continuous improvement tool as part of an asset management improvement programme.

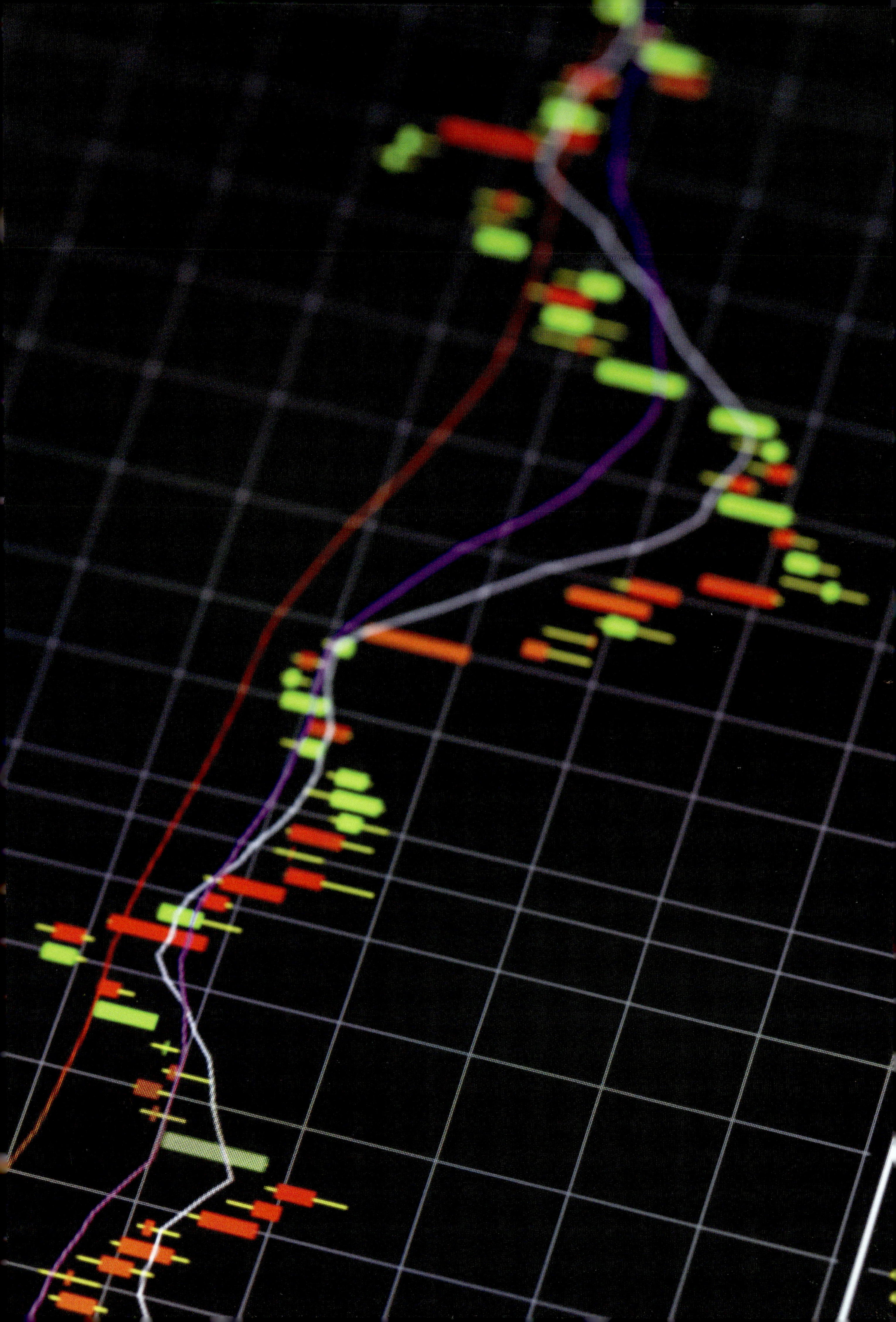

12 The organisation and the people dimension

Many of the factors that have a strong influence on effective asset information management relate to the culture, organisation and people that provide and utilise asset information.

12.1 People

12.1.1 Roles and responsibilities

It is important to understand the range of interests in information of various stakeholders, determined by their position in their organisation and their job role. The diagram below illustrates the broad categories of role involved:

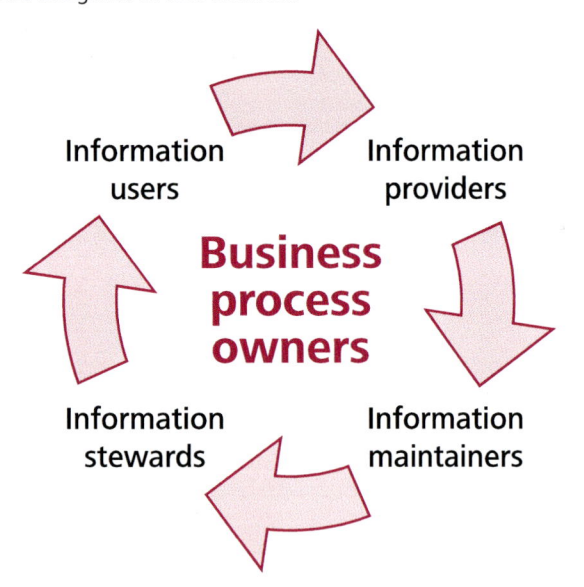

Figure 15 - Information roles and responsibilities

These typical roles are defined below.

Business process owners	Those accountable / responsible for defining the information required by a business process including its quality. There may be multiple business process owners for one set of information each with different requirements.
Information providers	Those responsible for providing information as a result of business activities, including analysts, designers, field staff, manufacturers and contractors. Information providers often have a very poor appreciation of the extent to which their information management activities impact on the quality of corporate information.
Information maintainers	Those responsible for validating information submitted from field staff or external contractors and physically updating the recorded information, for example drawing or records office staff.
Information stewards	Those responsible for supervision of information-related processes and functions, and day-to-day improvements to information quality.
Information users	Those using asset information for tactical, operational and strategic purposes. This may involve synthesising new information from existing information. Information users may be outside the organisation including regulators, partners and suppliers.

Information governance roles tend to cut across the organisational hierarchy rather than align with it. A presumption that all staff are involved in managing asset information quality will be closer to good practice than a presumption that it is only the responsibility of a few named individuals.

It should also be recognised that everyone can fulfil multiple roles at different times or even concurrently. For example:
- An information user may combine two existing information sets to synthesise new information or fill in information gaps by inference. They therefore act as information providers;
- Field staff may provide information on asset condition or performance but may use information on required maintenance activities partly based on that information;
- Field staff may directly enter information onto corporate systems using mobile technology; and
- A business process owner who is the only user of a set of information may also act as information steward and even information maintainer for that information.

If relevant asset activities are outsourced, then staff from those contractors will be stakeholders in asset information management. Suppliers, partners and regulators may also be stakeholders in asset information management.

Typically, the more knowledge of information structures and good practice asset information management that role holders possess (particularly Information Stewards), then the more likely they will be to identify and act appropriately on any information issues identified. The provision of better tools to access information will further enable and enhance this user driven improvement in information quality.

12 The organisation and the people dimension

There are also other support roles that are important in the management of information quality:

Information & systems support groups	Software specialists who are able to provide specific advice, manage information upload and quality improvement tools, and test and report on the quality of information, may also provide some front-line support for permissions and management of the "look and feel" of screens
IT support & projects	The IT organisation that provides the day to day IT service, infrastructure, communications and security, usually with the support of external suppliers.
Steering group & executive sponsor	Provides overall guidance to all information governance efforts including the championing of the governance strategy to all areas of the organisation.

> **Key point:**
> Although the term 'Data Owner' is used widely, many organisations find data ownership difficult to apply in practice due to data typically being used by many processes. Many organisations find it more effective to assign Process Owners who are responsible for ensuring that data used and created by the process meets organisational requirements.

12.1.2 Capability

In order for an organisation to manage asset information effectively (or any business activity!) there should be a clear understanding of the overall capability of the organisation. This should include both the skills/capabilities that it can utilise and the level of resource (include external suppliers) that could be utilised. Additionally, awareness of current and future demands on these resources will help identify potential shortfalls and bottlenecks.

Effective information initiatives require a combination of business knowledge, asset experience and IT skills so are therefore best undertaken by a blended team with access to all these capabilities. Factors to consider include:
- Experience of the whole life asset cycle can provide the expert oversight to get value from information cleansing initiatives;
- Over bureaucratic IT project approval processes and the commercial framework within which IT application support is provided can be a key constraint to delivery timescale;
- Knowledge of how processes and software use the information is also a key factor that must be well understood when managing information, particularly in developing automated identification or correction scripts;
- A staged process allowing reporting to peers provides better transparency across team capabilities;
- Business experts may range from the owners or stewards of the information and may extend to those involved in regulatory or wider business reporting functions;

- Information entry teams and site survey teams can be supplemented by hiring-in additional temporary staff. Due to the long term consequences of any information errors introduced, these resources are usually best employed in carrying out clearly defined and predictable tasks under strict governance and compliance processes. This approach frees up experienced resources to focus on more complex activities;
- Information analysis specialists, and those with expert knowledge of the applications whose information needs improving, can often make a major contribution to analysing the issues and helping to identify the most cost-effective solutions; and
- Stakeholder dialogue procedures offer the potential to draw in other experts in the business with vital knowledge of specific information or business areas. These staff may not have formal information responsibility but hold valuable experience from previous initiatives, whose contribution develops confidence in the outputs of the initiative.

These differing levels of capability in terms of understanding information, front-line asset management, and business awareness, as shown in Figure 16 below.

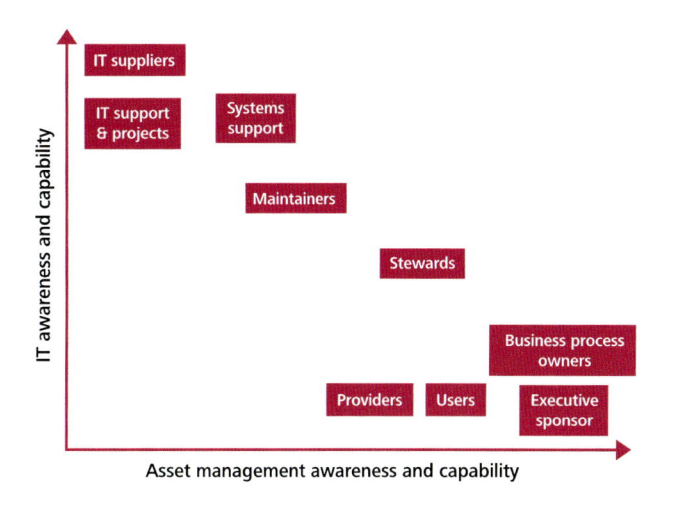

Figure 16 Capabilities and awareness for each role

It is vital that all parties are aware of the knowledge and skills each role in the spectrum brings to sustaining good information quality, even though they themselves do not and cannot posses all those skills. Those with high asset management and business awareness need to specify requirements that are then translated into increasingly technical IT terms within the support functions.

12.1.3 Culture and behaviours

It is important that a culture of information quality improvement is promoted which ensures that staff are aware of the importance of good quality information as part of normal day-to-day business activities. Improvements in information quality are delivered and maintained through the actions of individuals engaged with a common culture across a wide range across business functions. Engagement within the information culture is essential to make and sustain improvements.

An engagement strategy includes a series of approaches to focus individuals on the improvement of information. Tactics to develop such a culture include:

- Effective governance processes to support new business requirements which may require new/improved information processing activities;
- Awareness of roles in the overall process to encourage ownership of information quality within their influence;
- Raising awareness that everyone has a role to play in improving asset information, not just the Information Stewards;
- Clear allocation of information stewardship responsibilities within daily activities to promote reliable collection and use of information;
- Regular communication to staff, highlighting the benefits of good quality information and the impact of poor quality information;
- Reminders to staff of their own role in improving the quality of information;
- Identification of staff who are not delivering their information role and provision of training or mentoring to improve this;
- Inclusion of information quality aspects when setting individuals' performance objectives, and when appraising against these subsequently; and
- Recognition and reward of people who have improved approaches to information.

12.1.4 Training

Training is clearly necessary for new technology or processes, with refresher training providing reinforcement of objectives and learning points from practice. It is also important to consider how competence is measured in order to provide positive incentives and avoid unintentional consequences. While dedicated collection resources deliver greater reliability at a cost premium, routine information collection as part of day to day activities provides a sustainable route to information quality. Incentive and reward systems require great care as individuals soon realise how to achieve best reward with a minimum of effort, so supervisory costs must be included.

Users need to understand not only how to access the information they need for their routine work but what values really mean and how the quality of the information being used may affect decisions. They need to have a clear understanding of their responsibilities in providing all required data, to the right quality, as part of their normal role. A deeper understanding of information quality issues is particularly necessary for asset management staff. They need to understand the asset, the information describing it and the sources and accuracy of that information in order to develop effective asset management strategies.

Work on the asset base is often outsourced to contractors not only for new build projects, but also routine maintenance work. This means that the staff who actually collect information may not work for the organisation that owns the assets. In these cases, particular care is required to incorporate information return within procured contract deliverables. Compliance procedures must enforce the information return culture with active measures being often necessary to bring about a co-operative culture between the staff of the asset owner and those of the contractors. Robust and transferable information return processes underpin successful outsource initiatives.

Lessons from experience
Perverse incentives can occur if the information collector believes they will be rewarded by extra work opportunities if they exaggerate or miss-report.

Exception reports and information-quality dashboards remind users when their records are incorrect or incomplete. Scheduling a dashboard refresh and sending that information to managers is a great way to encourage compliance.

Performance targets can increase productivity at the expense of good data provision.

One utility company adjusted the amount of validation undertaken before uploading depending on the individual information collector's history of quality failures.

Case Study – Change communication
A major UK electricity distributor was engaged in a long term process to improve the quality of information within the business. This has been reinforced by a sustained communication programme to all staff which included: Letter from the CEO; Managers FAQ; Customer information journey poster; an 'Information Charter'; Information Policies Guidance with Supporting DVD; Training for 1800 staff. This generated 430 improvement ideas, delivered over an 18 month period and a recognised ongoing process.

12 The organisation and the people dimension

12.2 Key Messages

i. Clear roles and responsibilities relating to data should be defined. This can include clarifying the providers, maintainers, users and stewards of the information and the process owners that use/create data

ii. Governance should provide oversight of the specification and functioning of these roles

iii. Other groups such as Steering Groups, Projects and Support Groups can also help in the management of information quality

iv. Capability defines the activities that an organisation and the teams within it are able to execute

v. Competences refers to the specific skills that staff have gained, both through formal qualifications and experience

vi. Behaviours are how staff react to different situations when no-one may be observing them, which may be different to the way they behave if a manager/auditor is present

vii. Training should be used to provide, or reinforce, awareness of process activities, how to provide and use data and the overall approach to information management

13 Process

Most activities of an organisation should be specified and managed as part of an overall business process. Typically, data is an input to an activity, may be acquired or changed by that activity and then be an output from the activity. This output is then often an input into the next activity in an overall process.

In order for processes to operate effectively, it is important that they are mapped in order to show the different activities that comprise the process, how they interact and who is responsible for undertaking an activity (see Section 14.5.1 for more information). Process maps and detailed procedural steps should be easily available to all process stakeholders.

The process of defining and mapping a process also allows the information needs of the process to be determined. These data needs should become a core part of the process documentation. These data interactions should also be compared to the overall information requirements for asset management (detailed in Section 7). This should help to ensure that the data needs of the process are part of the overall information requirements for the organisation. If necessary, information needs should be revised in the light of the information needs of the process.

Effective process management typically involves the use of process monitoring to assess key performance metrics which allow opportunities for improvement to be identified. Included in these process metrics should be measures to reflect any information acquired or changed as part of the process which will help ensure that information is provided as intended.

The organisation should ensure that there is suitable governance established for the process and that these performance measures should be an input to the governance process. The governance body for a process may be the same as that for asset information, or it may be a separate group. In this latter case, it is important that effective communication exists between these groups to ensure that any poor performance, emerging business needs and issues to resolve are all effectively addressed.

13.1 Key Messages

i. Processes are clearly defined, mapped and followed
ii. The data inputs and outputs from process activities are clearly understood
iii. Critical data required to support business process is understood
iv. Process metrics defined and utilised appropriately
v. Process owners are in place to oversee both the running of the process and the data arising from the process

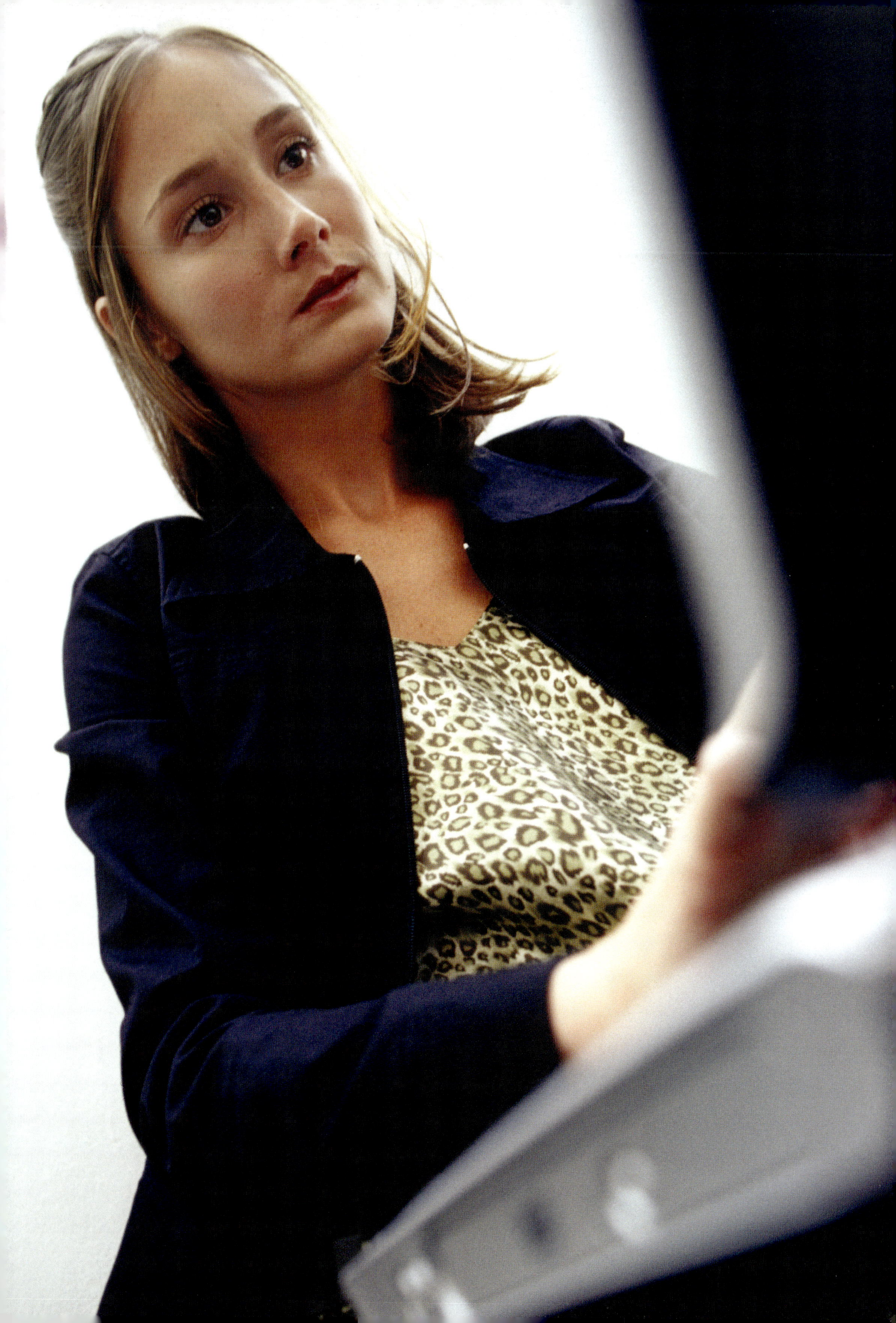

14 Software

In combination with people and business processes, appropriate use of information technology can make a significant difference in the way assets are managed. Software applications can store the data required for the organisation, and its asset management activities, to operate successfully. Software should automate processes, ensuring that staff are able to access appropriate information when it is needed which may also include remote, field based access to data. Appropriate automated interfaces between applications can ensure that, once a value has been entered, the same value is used wherever it is needed by other applications, thus avoiding the additional cost of repeated data entry and of course the risks of errors that that introduces.

The visual appeal, usability, stability and supportability of software can also have a significant impact on how effectively it is used, and hence on the quality of data.

Technology generally comes at a cost, and how beneficial it is to the business should be carefully evaluated:
- Do you need it now?
- Will another technological development in the next year or two be a more effective and cost-effective solution?
- Is this technology proven, or might it introduce an unacceptable risk?
- Technology has the potential to produce vast volumes of data. Do you need this volume and can you handle it?

Note:
The term IT (Information Technology) is often used both to mean the department controlling the technology and the technology itself. For clarity, in this document, when the department is being referred to, the term used will be "Information Technology department".

Note:
As the procurement and implementation of software solutions is a rapidly changing environment with many sources of good practice information, they will not be covered in this SSG. This Section of the SSG will focus on related activities that should be considered in order to secure more benefit from asset information.

14.1 Getting the best from information technology

When software is implemented and utilised effectively it can have a significant positive effect, both on business objectives and on ensuring that asset information is acquired and managed effectively. As such, effective use of information technology can differentiate world class organisations from poorly performing ones. It should be noted that, the effectiveness with which software is utilised is, arguably, far more important than the choice of software product itself.

Poorly implemented software, even if it is 'best of breed' enterprise class software can be worse than doing nothing. Wrongly implemented processing or system interfaces can easily result in slow and subtle data quality degradation, sometimes without the business users even being aware that this is occurring.

Example:
An electricity meter operator's systems were largely driven by request messages from the associated energy suppliers and the industry master registration systems. There were several types of these messages and these could be received in various different orders. Unfortunately not all orders of receipt had been correctly catered for in the software, so when requests arrived in those sequences they were not processed correctly and information was lost.

Errors can also arise if systems are not provided with all the information they need to make their decisions.

> **Example:**
> An electricity distributor used a geographical system to associate each customer premise to the nearest low voltage supply cable. Unfortunately, it had not been provided with the voltage levels at which the premises were fed, so it erroneously recorded high voltage consumers as being supplied from the low voltage network.

14.2 Enterprise Architecture

A number of important strategic activities which organisations would typically undertake as part of the management of data, business systems and processes are referred to under the heading of enterprise architecture. Although these activities are not specifically intended to resolve data quality issues, the rigour and analysis that they encompass can help to identify and quantify data quality issues and to identify possible solutions.

Large organisations may have autonomous units with strategic responsibilities; other enterprises may have more ad-hoc arrangements distributed through operational and tactical datasets. The aspects of enterprise architecture which are most relevant to asset information management are described in the following sections.

These techniques are used to both document current approaches and to document future planned positions. This then enables strategies for business improvement to be developed and implemented.

There are a number of Enterprise Architecture methodologies that are relevant to the development of an Asset Information Strategy which will not be repeated here. These include:
- Integrated Architecture Framework (IAF)
- MIKE2.0 (Method for an Integrated Knowledge Environment)
- MODAF United Kingdom Ministry of Defence Architectural Framework
- OBASHI (Ownership, Business Processes, Applications, Systems, Hardware, and Infrastructure)
- TOGAF (The Open Group Architecture Framework)
- Zachman Framework

14.3 Information Architecture and the Data Dictionary

An Information Architecture is a documented, business level definition of what data needs to be held by the organisation in order to fulfil its functions, and how this is achieved by the information landscape. The Asset Data Dictionary, as referred to in section 7.4, will typically be a sub-set of the overall information architecture for the organisation.

A complete information architecture can cover narrative explanations, diagrams and catalogues of:
- The entities about which information is held (e.g. customers, assets, documents, etc – including abstract entities such as locations, organisational units, jobs etc);
- The metadata that describes those entities and their attributes and relationships (see section 13.3.1 below);
- The taxonomy of these entities (e.g. that valves are a type of asset);
- The relationships that exist within the data between entities (e.g. an asset can have jobs performed on it; a substation may contain several circuit breakers);
- Which data is held in, and used by, which systems; and
- How data flows through and between the systems, and which interfaces are used to transfer it between which systems, including details of the exact data sent across each.

Depending on the particular needs of an organisation and resource constraints, it is likely that the information architecture will be developed as a number of prioritised phases over time.

The activities required to define the information architecture, including both business and technical metadata, will help to improve the data quality of an organisation by ensuring that there is common understanding of the intended meaning and use of data fields and of the rules required for validity checking. This process will also help identify areas of inconsistency.

A by-product of these activities which can help to improve the quality of asset information is a greater awareness of the structure, storage and uses of asset information. Detailed data models will help to identify cases where data is used repeatedly across an organisation. Additionally, greater awareness of the data that already exists can help prevent the initiation of data gathering activities to obtain data that is actually already stored somewhere else, thereby allowing more time to be spent on more productive data quality improvement activities.

14.3.1 Data Dictionary management

An important major constituent of an Information Architecture is the Data Dictionary, usually defined as a repository of metadata ("data about data"). This information will typically be stored and published in metadata repositories and should be supported by relevant data models and dictionaries. Data Dictionary information typically falls into two main categories:

Business metadata:
- Definitions of the meanings of data fields (for example the field SerialNo represents the Serial Number of an asset and the field Condition represents the condition grade of an asset)

- The format and coding conventions used for the primary identifiers (e.g. asset numbers, works location codes) by which each entity instance can be referenced by other items of data; and
- Their sources and (ideally) usages.

Technical metadata:
- Definitions of the structure of data that is primarily used by database administrators (for example the field SerialNo is a 14-character text field which may not be blank, and Condition is a numeric field containing the ID of a value from table CondGrades with a default value of Unknown);
- Which field(s) form the primary key (unique identifier) for each type of record (e.g. asset IDs, job numbers, etc)
- Which fields hold primary identifier values of other, related records (e.g. the ResponsibleAssetID of a job).

If the quality of the existing data is known to be imperfect, and this is not expected to be resolved in a relatively short time via a data cleansing initiative, it can also be useful for the data dictionary to include fields where the quality of each specific item can be recorded. This could be as simple as a free-text Quality field against each entity and field in the dictionary, or could be extended further to include metrics such as the (exact or approximate) percentages or records that actually have the field(s) in question populated.

14.4 Data Mastering Strategy

A Data Mastering Strategy is a methodology that defines which data stores are regarded as the 'system of record', i.e. holds the master version of, each element of the Information Architecture. It will also explain how the copies used by other applications are accurately maintained, for example via interfaces between the applications. This approach is typically taken where a considerable number of data items are needed by several different software applications, all of which could otherwise contain different data for the same entity. It helps the organisation determine which application should hold the master of each piece of information, and thus where updates to it will be managed.

In practice, it is extremely common for assets like cables and pipes to need representing in more than one application, and by no means unknown for there to be as many as four or more applications that need to hold them. For example a Geographical Information System (GIS) may be used to record asset locations whilst an Asset Management System (AMS) tracks their manufacture details, condition and maintenance records. As these types of software each have specialised features, and off-the shelf software products that fulfil these differing roles are available, it rarely makes sense to try combining everything into one single bespoke application. As soon as the same data item is needed in more than one place, the question of how to manage those copies arises.

The term duplication of data refers to where a piece of information is stored in two or more data stores with no automated processing to ensure the copies are maintained consistently with one another. This is considered bad practice and should be avoided due to the high likelihood that these data stores will not agree with each other.

Replication of data refers to where data held in one data store is automatically copied to other data stores in a manner which ensures continued alignment of those data stores. Replication of data is frequently employed within application landscapes in order to meet performance and availability requirements, or to provide data in a form needed by Commercial Off-The-Shelf (COTS) software products to perform their functions. Replication processes should be carefully designed and managed to ensure that they can correctly process all combinations of data and of the order in which events to it occur.

Any data store could be treated as owning the master copy of a particular data item, with new values being entered into that store first then replicated to the others by automated interfaces. Depending on the organisation's business processes, or other requirements, it may be better for one particular application to be nominated as the master, and the same answer may not be appropriate for different organisations. In some cases it can even be appropriate to master some attributes of an asset in one application and others in another, for example its geographical location in a GIS but its manufacture, age and condition data in an AMS.

Whatever the conclusion, however, the important points are to ensure that the strategy is thoroughly reviewed and agreed, and then clearly documented and enforced as a key facet of information governance. The strategy should ideally also be made easily accessible to both business and IT department staff. Data Quality can be improved by making sources and mastership of data more transparent in this way because it arms staff with information they can use in day-to-day decision making.

14.4.1 Master Data Management

Master Data Management (MDM) solutions are specialist technologies designed to support and automate a Data Mastering Strategy. When appropriately deployed, following careful cost-benefit assessment, this software can automate many routine tasks involved with improving and sustaining asset data quality, and provide a toolbox to enable data stewards to configure additional checks and monitoring when appropriate.

MDM technologies are designed to help organisations maintain their master data and, in particular, make available a 'golden

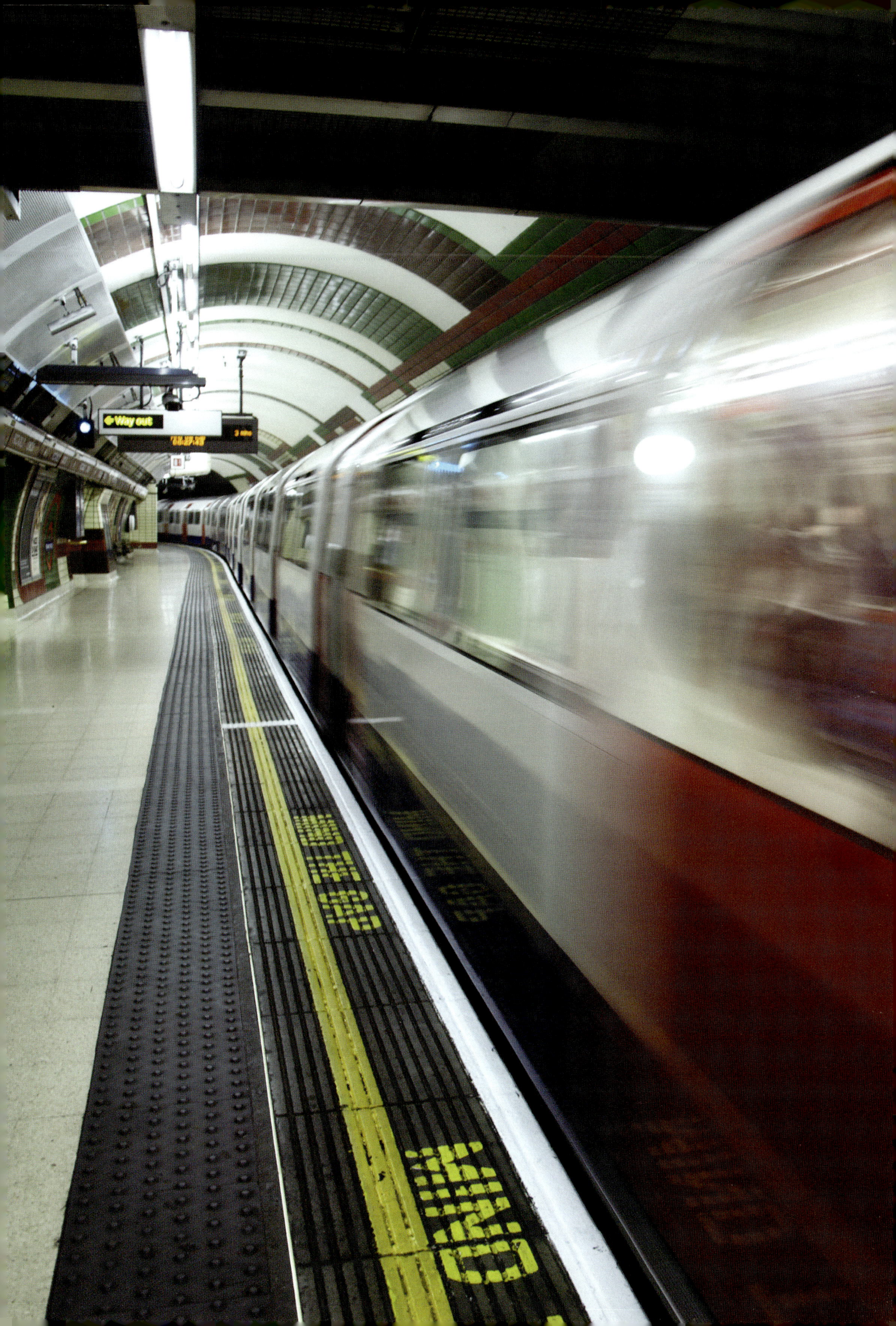
Way out

record' of each data item which encapsulates a single version of the truth. This can either be physical, with a new central repository built to hold the master copy of the data, or federated, with the common view derived from data mastered in individual systems. The specialised nature of systems commonly used with managing assets tends to favour the federated approach when applying MDM technologies to asset data.

14.4.2 Data profiling/ Data Quality Management tools

Data profiling and Data Quality Management (DQM) technologies started out as toolboxes of powerful data profiling, analysis and manipulation routines. These include facilities for comparing two (or more) lists and reporting on items that appear to be missing from one or more of them, e.g. for locating mismatches between asset datasets that have not been managed ideally in the past. They can also help with compiling or managing data dictionaries as they can not only read the technical metadata directly from the databases themselves but also rapidly compute key quality metrics such as the numbers of records that have values in a particular field.

There has been considerable convergence between MDM and Data Quality Management facilities, and often both are included within product suites. Both allow business data quality rules to be defined, for example "any asset with a status of Commissioned must have a valid Location ID". They can then monitor the data and produce regular reports of any records that violate any of the rules defined. The lists of transgressor items can automatically be transmitted to data stewards or other relevant staff for remediation. They can also produce dashboard reports configured with relevant quality metrics for the asset data as a whole.

14.5 The role of information technology in projects

It should be noted that it is rare for there actually to be an "IT Project" – projects that acquire this label are typically business change projects that are supported by information technology activities. The technology part of a project typically goes through a set of standard stages from conception through to delivery and implementation. There are two classic kinds of project lifecycle:
- Sequential lifecycles, sometimes referred to as the "V" or "waterfall" model, in which the project progresses through stages such as initiation, definition, production and implementation. The early stages concentrate on producing a detailed specification of what will be done, then the later stages carry out those instructions.
- Iterative, or "agile", methods in which a basic version of the solution is prepared within a sandpit environment and then successively refined until an acceptable version is achieved.

Sequential lifecycles have the advantage, at least in theory, that when the specification has been sufficiently defined the remaining work can be contracted on a fixed-price basis, so are more frequently employed than agile methods. When using sequential lifecycles, all system requirements need to be captured during the initial stages, so it's important that sufficient asset management and data experts to support this are made available to the project right from the start. The down side of such projects is that they can be inflexible if business requirements change, particularly for projects running over long time periods.

Agile methodologies have traditionally been used for simpler technology solutions, however, they are increasingly being used for enterprise solutions.

Projects are delivered in accordance with a project governance framework which typically incorporates defined stages. 'Gate reviews' are undertaken between consecutive stages, to review the products of the preceding stage and give approval for work on the next stage to commence. Stage gate reviews should be involve asset management staff including asset data stewards. Various standard project governance frameworks exist, including PRINCE2 (projects) or MSP (programmes, which should be utilised to minimise the risk of project failure.

System integrators work under considerable cost and time pressures, and projects may involve compromises on functionality or quality as a result. Successful projects need considerable levels of business ownership and input. Unless the asset management function remains closely involved as software is implemented, and the developers are provided with full and appropriate sets of system and implementation requirements, they will not have the necessary knowledge and direction to guide them to the right choices. When IT projects or programmes are undertaken, effective collaboration between the business and IT functions really can mean the difference between mediocrity or failure and success. See Section 15 for more information on Managing Change.

14.5.1 Business Process Analysis

Business Process Analysis is where the (old and new) processes by which the organisation manages its assets are identified, agreed and documented. Depending on the scope of the project, this could include processes for investment planning, inspections and maintenance, capacity planning, day-to-day operations and emergency repairs, and/or disposals.

If the existing data is known to be imperfect, processes specifically for data management, such as addressing data issues as and when they are identified, should also be established. The system requirements for supporting these processes can then be appropriately determined.

It helps considerably if the processes are defined to a level of detail that includes descriptions of which data items are created, used, modified and deleted at each stage in each process. This gives everyone much greater visibility and understanding of the end-to-end data handling requirements,

14.5.2 Requirements Definition
Unless a so-called "agile" methodology is used (see above), there will be a phase early in the project in which a set of detailed requirements for the new system is compiled. Input from the asset management function is required to provide detailed input as to the data processing requirements, and the current and targeted levels of quality.

Including a specific section of data management requirements can be a very good way of focusing the project on supporting good data quality. Where appropriate, consider including:
- a requirement that software is designed to ensure data quality is not degraded, and if possible is enhanced, by its implementation;
- requirements for data to be validated as comprehensively as possible at its point of entry;
- the ability for data stewards to view detailed logs of all historical changes made to specific data records;
- requirements for producing detailed logs of any errors the system or its interfaces report with enough detail to identify underlying causes and data issues ;
- requirements for implementing processes and workflow by which those errors can be resolved; and
- a requirement for the project to update the organisation's Information Architecture and data quality knowledge base appropriately prior to demobilisation.

If it is known that the quality of the data is less than ideal at the outset, it can also be appropriate to require that relevant automated interfaces include mechanisms whereby re-runs or refreshes can be triggered manually where required, for example after a batch of intensive cleansing has been done.

Detailing such points in the Requirements Specification can be a major step forward for data quality. It draws a line in the sand beyond which the project cannot then step without raising change proposals for wider discussion amongst its stakeholders. It can be made more easily repeatable from one project to the next by producing a template containing generically worded requirements which can then be included in the each new one's Requirement Specifications after appropriate customisation.

14.5.3 Business Case
All projects need a suitable Business Case to detail what is planned to be achieved, the costs of the project and the anticipated benefits. If the cost/benefit ratio is too low, then this indicates that the project probably should not proceed. As a project progresses, new information to support firmer cost estimates and benefits should be included. The cost/benefit ratio should be reviewed at each project stage gate to confirm the desirability of continuing the project. See Section 15.4 for more information.

14.5.4 Data Migration
The final area that needs considerable asset management team involvement is data migration to move data from the source system(s) to the target system(s). Statistics on data migration projects indicates that they can have a high failure rate, particularly if this is treat as an IT department activity. The use of formal data migration methodologies, such as Practical Data Migration version 2 (PDMv2) are credited with reducing the failure rate of projects significantly.

Data migration should be treated as a business project supported by the IT Department. This is Golden Rule 1 of PDMv2. Other key aspects of a successful data migration that the methodology embraces include:
- Key data stakeholder management – Ensuring that senior business and relevant data stakeholders are kept informed of the migration project and fully engaged in decision making – an essential for success
- Migration analyst – an experienced analyst with good business and data skills will be needed for all but the simplest data migration
- Landscape analysis – Exploration and discovery of legacy data, also called analysis and profiling, is a key activity which should be commenced as early as possible. If this is commenced at the outset it is sometimes possible to influence the detailed specification of the new system(s) so it will better handle expected data quality levels
- Gap analysis and mapping – This is where the legacy data is mapped to the new system(s) and any gaps identified are fed into the data quality rules process below
- Data quality rules – These cover the process of identifying and agreeing how any data quality issues or other incompatibilities with the new system(s) will be resolved, for example, by data cleansing. This is managed as an ongoing process, with the full engagement of the business stakeholders a key requirement for success
- Legacy decommissioning/ System retirement plans – Defining the criteria on which the business are prepared to approve the decommissioning of the legacy systems is a key method of ensuring that data issues are appropriately addressed
- Migration design and execution – The design, test and execution of migration and archiving legacy data. Data which fails testing indicates the need to add/amend data quality rules
- Migration strategy and governance – Programme management and governance of the migration project

Where possible, early trial runs of the migration process are recommended so that any remaining issues can be identified and fed into the data quality rules process.

Where data quality issues occur in the context of the data migration:
- Some additional cleansing options become available, namely 'on-the-fly' during the migration process, or afterwards in the new system(s); and
- Legacy system(s) may not provide any easy means for data to be corrected in them prior to the migration

The completed migration will need testing by the appropriate business users/ data experts.

Example:
A major project's data migration was reporting eight-figure numbers of errors and only very belatedly was a data analyst called in to investigate. No data profiling or discovery had been done so he addressed this. When a summary of the data patterns present in a coded location ID field was extracted into a spreadsheet and given to the data steward, an initiative from a few years previously when a new scheme had been trialled in one district came flooding back into his memory. Until this point it had lain completely forgotten so the migration designers had had no reason to cater for it in their conversion software. Some details of the new system's design then had to be re-worked as they had also been based originally on the false assumption that the field always contained different types of value.

Lessons from experience:
Bad data is typically listed as one of the top three reasons why projects fail. Bad data leads to misleading, incomplete, and confusing information, resulting in a loss of trust. It doesn't matter what type of project you are attempting, once a perception that data quality is poor has been established, it can be a very difficult perception to correct.

Example:
It is common to underestimate the effort needed for data migration. Many asset managers have had experiences of important data missing after a change of asset management technology or of data available but no longer easily extracted for analysis using separate software.

14.6 Key Messages

i. Technology can assist greatly with improving and sustaining data quality, but can be expensive and should not be deployed without a satisfactory business case.
ii. Successful IT projects or programs can only be achieved through good collaboration between the business and the IT function, both of whom must play their allotted parts.
iii. Business process analysis must cover all relevant activities and highlight which data is accessed at each stage.
iv. Clear requirements should be established for IT projects to ensure that they don't degrade data and provide enablers for ongoing data improvement.
v. For projects or programmes, analysis of existing data should be commenced at the earliest opportunity.
vi. Data migration should be led by the business and should not be under-estimated.
vii. Appropriate management of the issues uncovered by analysis and during migration is crucial.
viii. A clearly defined and well-documented information architecture helps everyone understand the intricacies of the data and its inter-relationships.
ix. The data dictionary can also become a useful vehicle for recording known data quality shortcomings.
x. A clear data mastering strategy is a crucial part of an Enterprise Architecture.
xi. Master Data Management (MDM) can be employed where data mastering strategies are more complex.
xii. MDM/DQM solutions can also help improve and maintain data quality by monitoring the data, reporting discrepancies to appropriate staff, and supporting their resolution.

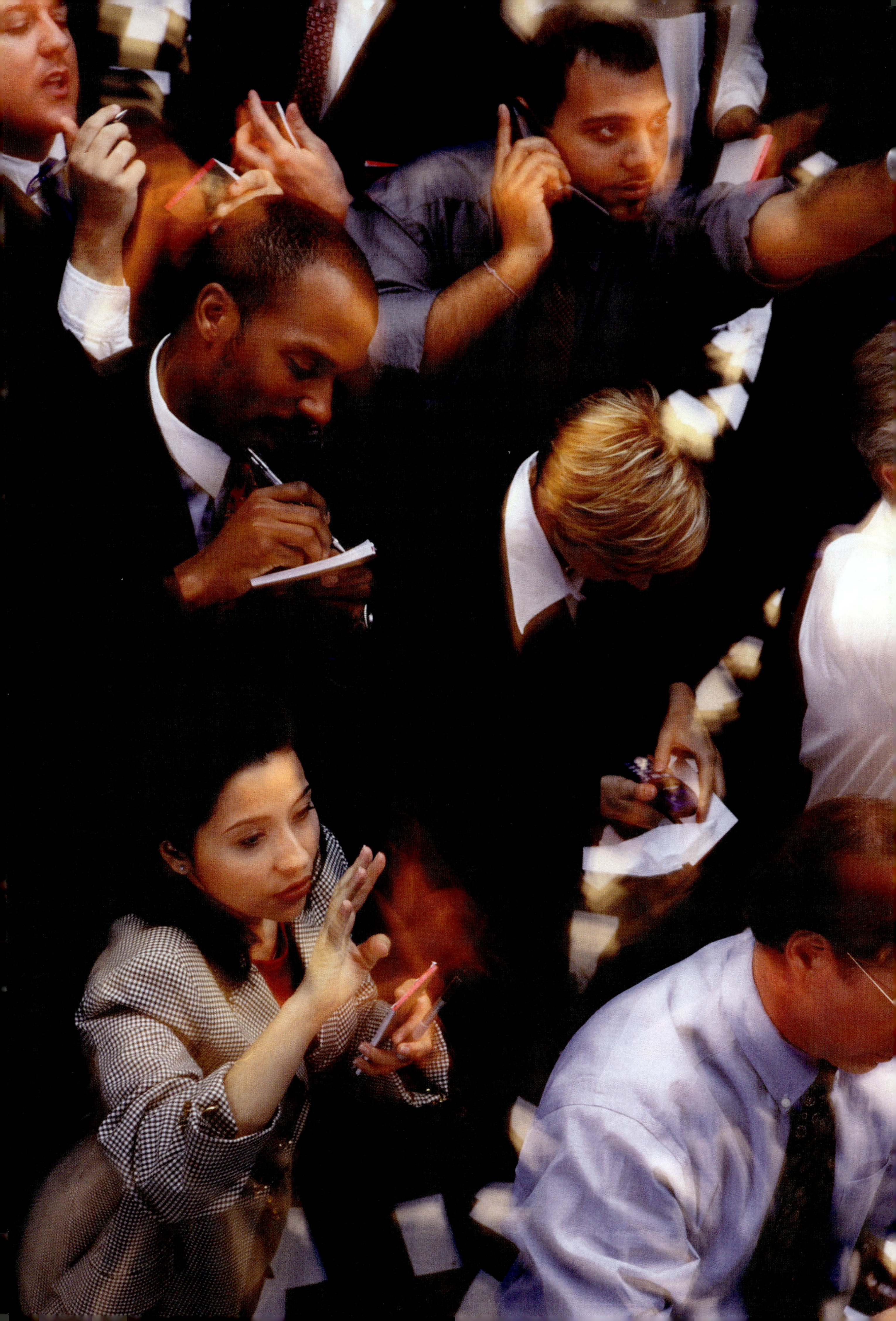

15 Managing change

15.1 Introduction

Changes to asset information need to be built on an understanding of the current state of information in the business (see Section 8.1.1), the needs of all asset information users and the "toolbox" of options available for making improvements to data. There is inevitably an iterative process as discussions with information users and providers will inform the understanding of the possibilities and needs for changes and the costs and risks of different options.

At this point, the organisation has assessed what asset information it needs, related that to its business drivers, and to what information it actually has. It now needs to plan a business change initiative to:
- Fill the current information gaps. This could be a project or series of projects to define the structure to manage this data, to create the IT systems or business processes to support this, and to collect and validate the data.
- Maintain its defined asset information base, on an ongoing basis. This could be a culture change programme, to demonstrate the need for the information to be maintained, and to instil in all relevant parties the culture of doing so.

There are three different situations to which this may apply:

15.1.1 Green field situation
Where the organisation is being started from new, there are no "gaps" as such to fill, and the objective is to create the optimum asset information structure from the outset, and to ensure contractually and culturally that manufacturers and service providers support the new asset infrastructure with full and accurate information.

15.1.2 Major business change
The trigger for change could be as a result of a merger, acquisition or major management change. The organisation is facing a situation where the existing processes, technology or scope require revision. In the case of a merger, this would be the bringing together of two (or more) separate sets of processes and systems. Alternatively, there may be an initiative for a step change in the performance of the business.

15.1.3 Improve the quality of existing asset information "systems"
For example, where requirements and organisational maturity have changed so that the organisation needs more data, and

also is now able to produce and manage it. A similar scenario would be where the system which has served the organisation now no longer functions as effectively as it did, because of changes of processes, people, technology or the assets themselves.

In managing a change in asset information management systems an organisation will need to build on the detailed considerations that have been set out in the preceding sections. This section provides an overview of the change process and flags links to relevant detail that should be considered from the preceding sections.

15.2 What is to change

Asset information comes from "systems". These are the systems of work which cause information to be captured, updated and distributed. They include:
- The process of managing asset data through the business. (Who collects it, who checks it, the route it takes, and whether it is stored centrally or distributed);
- People, training, motivation for the tasks that they are expected to do;
- Software systems (corporate data systems, interfaces, access to data by third parties); and
- Data collection technology which captures information from operational and maintenance processes (this may include automation, mobile workers with hand-held devices, etc).

Additionally, substantial checking and improvement of data may be made by using logical assessments of data for consistency (with related data), extrapolation, interpolation, inferencing from related data etc.

Section 7 considered what information was required to manage assets in the desired way. This should be considered in line with the Asset Information Policy and Strategy. Section 8.1 describes assessing current data as part of the Asset information lifecycle and Section 11 describes benchmarking asset information management. The business must now assess what information is practical and cost-effective to collect. Section 12 highlights the importance of people in both on-going information management and in any change process.

15 Managing change

Consider options to collect and organise the essential information:
- Is the gap in requirements small, so that it can be covered by small changes from present processes?
- Does the gap arise from the data itself, the way it is processed or who requires it? (in or outside the organisation, for example the supply chain, the regulator);
- Would new data collection technology help?
- Is human knowledge an essential part of the process?
- What data collection and management systems are suitable and available, and what degree of integration is appropriate?
- Is data continuity an issue? Must metrics and trends be maintained from the old situation to the new one?
- What do similar organisations (in this or a parallel sector) do?
- In the light of the possible options, is it necessary to reconsider the level of data to be collected and processed?
- Various data improvement options are discussed in Section(6) above as part of the consideration of Asset Information lifecycle and many possible improvement ideas and their benefits and risks are discussed in that section.

15.3 Technology Considerations

In combination with people and business processes, changes of technology can make a significant difference in the way assets are managed. This is one aspect of the process of identifying options. However, in an asset-intensive business, with geographically-distributed assets, technology can have a significant contribution in the automatic collection of data from assets and in communicating information.

For example, mobile communications enables the collection of data into the business, and distribution of job and supporting information into the field. Data can be collected automatically in the field with greater speed, regularity and volume. Another example is that a railway route can now be surveyed accurately for track alignment, loading gauge, tree incursion, trackside equipment inspection etc at high speed from a moving train.

Guidance on considering the cost/benefit case of technology investments was given in Section 8.2.10.

15.4 Assessing Options and creating a Business Case

Once options for change have been listed initially, the likely approaches should be shortlisted. These should then be defined in more detail and outline "process designs" defined, in terms of their costs, benefits and risks.

15.4.1 Costs
- Business process changes incur a cost - This may not be a transparent cost, and may incur the use of time and resources hidden within the business. There may be a loss of focus and consumption of the time of management and key resources.
- Training costs - Almost any new solution will require some degree of staff training and the absorption of a loss of efficiency in the interim.
- CAPEX / OPEX tradeoff - Typically, a more technologically-based solution will involve more capital cost, but have a lower operating cost than a human-based solution. The choice between these should follow the financial objectives and guidelines of the business.
- Other system costs - Software systems have not only their initial purchase cost, but other costs, related to application and IT infrastructure support, licensing, configuring business rules and data, creating data structures, populating and migrating data, data quality and validation.
- System interface costs - The actual implementation of interfaces between systems can often be underestimated, and can prove more complex and costly than envisaged. Even where a published interface exists, this must be configured and tested.
- Ongoing support of business systems and incremental changes arising from business requirements may be done in house, or may require support from the supplier, which will affect the support costs.

15.4.2 Benefits
- Better decision making, through having the right information available to the right people. This may arise from a better understanding of the asset characteristics and performance, more effective processing and presentation of the data.
- More efficient processes, through the elimination of unnecessary or obsolete process steps, less duplication of data handling, no reworking of data. Other activities, which may be improved include needing to analyse data, locate or re-create data, verify data, which are typically ad-hoc, manual processes.
- Better ability to comply with, and prove compliance with, regulatory or legislative requirements.

Example:
a utility company may collect additional information on asset condition, which enables them to refine asset replacement decisions to reduce unexpected failures, and to avoid unnecessary premature replacement of assets. If applied to large volumes of low-cost assets, this could save considerable expenditure.

Subject Specific Guidelines: Asset Management

15.4.3 Risks

- An asset-centric business delivers services through the functioning of its assets, and the knowledge of their performance and condition. Failures in these can affect the delivery of service, and the reputation of the business, therefore the implementation of the project must always support "business as usual".
- Risk of aiming to create the perfect solution, which proves impractical to deliver in the required timescale
- A business developed a process using separate best of breed financial, asset and geographic information systems, however the interfacing required proved to be too difficult to complete. This led to the unpalatable choice of changing the original choice to another solution.
- Designing an effective solution, but without support ("buy-in") from key players in its delivery, and therefore the solution is doomed to failure.
- Risk of a mismatch between the users' expectation of the capability of a system and the actual capability when configured and applied to their situation.
- Getting carried away with enthusiasm, and failing to relate this to the reality.
- Focusing on delivering the change, but failing to think through the long-term support and maintenance of the processes, system and data. This leads to a progressive degradation of the effectiveness of the system, and gradual loss of trust in it.

The organisation should use this type of approach in building a business case. Key factors in the success of developing and selling the business case are:

- A clearly-articulated "Business Vision"
- An influential sponsor (a board-level sponsor may often be appropriate)
- A champion within the business who has detailed knowledge of the impacts and is responsible for its delivery.
- Stakeholder engagement, which must start at a suitably early stage, and be continued throughout the implementation. It gives them a chance to influence important aspects of the programme, and to "buy in" to it, through a sense of "what's in this for me?"

These factors all contribute to building "hearts and minds" commitment at all levels. Lack of these has caused otherwise good schemes to fail.

All projects need a suitable Business Case to detail what is planned to be achieved, the costs of doing this and the anticipated benefits. If the cost/benefit ratio is too low, then this indicates that the project probably should not proceed. As a project progresses, new information to support firmer cost estimates and benefits should be included. The cost/benefit ratio should be reviewed at each project stage gate to confirm the desirability of continuing the project.

For details refer to Section 8.2 which provides many examples of options to improve data and of the value of utilising data.

15.5 Preparing for the Change Programme

Asset information improvement is a change programme, affecting people, processes and systems. Once the basic concept has been agreed, it is important to look for effective ways to make changes in all three. This requires good process design and effective data management (if this is required for the selected approach). It also requires a plan to implement, carry through and sustain changes of human behaviour.

15.5.1 People

Having the right team of people will make a real difference to the successful implementation of the programme. This may involve in-house staff, a contractor and other consultants to facilitate and advise on the process. Particularly in demand will be the most skilled and knowledgeable resources in the organisation, as they will be essential to the success of this programme, yet will also be required for their regular operational duties. It is crucial that the operational staff who are selected for secondment to the programme are experienced in the particular areas needed, rather than simply available.

15.5.2 Detailed specification

Important aspects of the specification must now be thought through in more detail. For example, to ensure that the data is cost-effective and not under/over-specified, it is essential to understand:

- Who needs to use the data and why?
- Data ownership (is there one person who takes responsibility for each item of data, even if others use it?) For example, one person for financial and another for technical aspects of data / one person is responsible for mechanical asset data, and another for electronic asset data? / A management structure, such as northern region and southern region;
- Data mastering strategy (leading to a "single source of truth"). The rules eliminate ambiguity over sources and time (see Section 14.4); and

For details refer to Sections 5 and 6 which describe the requirements for specifying information governance and strategy,

15.5.3 Interfaces

The purpose of asset information is to support a wide range of facets across the business, and having clarity on the interfaces is essential. There will be interfaces between people and systems, between departments, and in many cases, with the upward and/or downward supply chain. Designing effective interfaces to capture information in one place, move it to other places

15 Managing change

where it is required, secure it and share it, is essential to success. Therefore, the interfaces – whether an internal network, the internet or a piece of paper – must be designed effectively, understood by the parties using them, and tested to ensure that the intent works in practice. At the outset, this must be recognised, expected and planned for.

For details refer to Sections 13 and 14 which describe the information process and systems

15.6 Programme Management

Appropriate programme management is essential. This should typically conform to a recognised standard quality system. The controlling document set should be clearly communicated to those involved. For major projects, there is a risk of wasting huge sums of money, and causing highly-disruptive delays, unless programme management is experienced and strong.

For major changes, involving several functions of the business, the programme management team should involve all the functions, and the programme manager should come from the appropriate part of the business, (which might be operations, asset management, IT department, etc). This team should be appointed early in the process to effectively plan and control the programme.

The business staff must interface with the IT department (if this is a technology solution) to implement progressive new systems, whilst both business and the IT department must manage day-to-day delivery of operational activities. Part of the management of the programme is to plan ahead for resource or business conflicts. It should recognise the critical or busy times in the calendar of business-as-usual, and make sure that the key points in the programme avoid unnecessary conflicts with these.

15.7 Creating the new Processes and Systems

15.7.1 Process Changes
Technically, it is often beneficial to prototype a process, to confirm that requirements have been fully captured and understood. This often has the effect of clarifying users' own understanding and their true requirements. This can take the form of "whiteboarding" the process in an intense series of workshops, or could use a process design office with charts pinned to the walls.

It is strongly recommended to make the design highly visible, and to test it with experienced users, to ensure that it is robust. Equally, it is important to have an independent view to consider whether users are raising comment from practical knowledge or merely "blind preference" of current practice.

Depending on the scope of the project, this could include processes for investment planning, inspections and maintenance, capacity planning, day-to-day operations and emergency repairs, and/or disposals.

If the existing data is known to be imperfect, processes specifically for data management, such as addressing data issues as and when they are identified, should also be established. The system requirements for supporting these processes can then be appropriately determined.

It helps considerably if the processes are defined to a level of detail that includes descriptions of which data items are created, used, modified and deleted at each stage in each process. This gives everyone much greater visibility and understanding of the end-to-end data handling requirements,

15.7.2 Testing
There should always be a Test Plan which sets out the process design tests required to formally accept the system / element. This should attempt to simulate a real set of the relevant business processes, but should not try to be so all-encompassing that it becomes impractical to progress through the testing. Knowledge of best practice in test design should set an appropriate level, for individual systems and subsystems, within the context of the whole asset management infrastructure.

New systems' user acceptance tests should ideally be conducted using actual migrated data and include runs-through of actual business scenarios, to give maximum confidence that the new system works correctly using the migrated data. All forms of new process should be tested before being committed to the live business, to validate the people and process aspects as well as technology. If this involves an IT solution, once it has been created for one part of the new asset information infrastructure, it should be tested in isolation, and then across its interfaces in an off-line environment initially, to avoid impacts on business as usual. This may have practical considerations, such as whether a test environment is available, both for the organisation itself, and for its partners.

15.8 Commissioning and changes to Processes and Systems

As new systems (software or other systems) are implemented, there is a process of commissioning the technology. In parallel, there may be changes to the processes which are used to generate and maintain data. Users must be trained in good time to be familiar with the new software and processes, but not so far in advance that they are able to forget the training. This means that if there is a delay in change implementation, this may require brief refresher training.

The programme plan will define a point at which the software is commissioned and goes into maintenance/support mode. It would, however, be advisable to allow for a period of dedicated user support which supports the users embedding the new processes. There may be support requirements arising from the need to access legacy data, or from migration issues.

The changes mean that people may be less efficient for a period, while they acclimatise to the new processes, and also while problems are identified, raised and rectified. This typically means that extra resources are required for a period to accommodate a "bulge" of activity, until the new process is accepted as the norm, and any backlog cleared.

During this period, there is higher than usual potential for things to go wrong. It is advisable to include an audit function which tracks data and changes to the data. This enables the organisation to see where, and when, things went wrong, and hence to rectify process problems and the training of people. It may also be considered whether this audit process should become the norm, to support continuing process improvements.

15.9 Culture Change and Change Management

In many cases, changes to human behaviour and activities are the key to the success or failure of the programme. The change programme needs to make good use of the sponsor and any business champions. There also must be an organised communication process, and a training plan.

Even when users have been appropriately trained, it is also expected that the business itself will learn lessons about its processes and data during the course of implementation, which will lead to changes in requirements. The Programme team should plan for this, and expect to manage it. They should therefore focus on finding the most appropriate solution, which may pragmatically make progress, rather than always striving to achieve perfection.

Where change is introduced into a large programme in an uncontrolled manner, there is great risk of causing increases in cost, and adverse delivery. Change should be raised formally and documented in a change management system. This allows all parties to consider whether they can accommodate the change, the impact on programme, additional costs and other matters. Once items have been raised, the programme team should assess the various items in relation to overall progress and only allow those changes which are essential or broadly beneficial.

For details refer to Section 12 which describes the people aspects of information management.

15.10 Key Messages

i. Decide whether you are starting from a green-field situation where you can design the optimum system, or from an inherited base of asset information, which may be inappropriate to the needs of the current business. Recognise the impact in terms of quality of existing data and embedded attitudes.

ii. Assess what information is required, and in what ways this information can be gathered. Is this realistic, practicable and cost-effective? If not, what alternatives are there which would serve the business drivers instead?

iii. Define options, assess likely approaches, and shortlist the most effective ones. Consider costs, benefits, risks and practical considerations. Prepare a business case for the preferred (most effective and most advantageous) option. Include the impacts on the entire business as well as the benefits.

iv. Maintain consultations with information users and data providers to ensure that the all appropriate change options are being considered and that costs, risks and benefits are fully evaluated.

v. Have a senior Business Sponsor, who holds and promulgates the Business Vision for the asset information programme.

vi. Engage stakeholders early in the process and respond positively to their input.

vii. Make sure that the programme includes experienced business users.

viii. Design conceptually the new processes and interfaces. Plan to communicate with those involved and to test effectiveness.

ix. Prepare a set of underpinning documents, and ensure that everyone is aware of them and works to them.

x. Appoint a team promptly, and engage sufficient experienced resource.

xi. Create the processes, implement and test the systems to a reasonable level.

xii. Plan the migration of data, and the retention of essential legacy data.

xiii. Plan for commissioning of the systems and the training and support of people through the start of new processes.

xiv. Expect a culture change programme

xv. Expect changes in requirements and the processes to manage these

16 Glossary of terms

Asset attributes
Details of assets, such as size, dimensions, serial number, age etc.;

Asset financial information
Purchase costs, running costs, valuation and disposal costs;

Asset inventory
A catalogue of all assets relevant to an organisation;

Asset location
Details of the location and spatial extent of an asset;

Asset management
Systematic and coordinated activities through which an organization optimally and sustainably manages its assets and asset systems, … over their life cycles for the purpose of achieving its organisational strategic plan;

BIM
Building Information Modelling/Management - a process involving the generation and management of digital representations of physical and functional characteristics of a facility

CAD
Computer Aided Design - the use of computer systems to assist in the creation, modification, analysis, or optimization of a design

CMMS
Computerised Maintenance Management System, similar to an EAM system, but likely to be smaller in scale

Condition data
Assessment of the physical condition of an asset independently of its performance

COTS
Commercial Off The Shelf – a description applied to software which is purchased by an organisation and implemented without customisation or bespoke changes

Data
Numbers, words, symbols, pictures, etc. without context or meaning, i.e. data in a raw format, e.g. 25 metres;

Design data
Information about the physical layout, performance requirements and capabilities of a facility. Can include 2D, 3D, 4D (Time/construction sequence) and 5D (Cost information)

Document based information
Drawings, documents, photographs of assets;

EAM
Enterprise Asset Management, a label applied to larger, more sophisticated asset management ERP systems, see CMMS

ERP
Enterprise Resource Planning, a label applied to complex proprietary systems typically used to automate some or all business functions of an organisation

GIS
Geographical Information System, a generic name for computer systems which store and analyse data using spatial information and relating this to underlying map information

Information
A collection of data expressed with a supporting context e.g. The span of the bridge is 25 metres;

Information Management
The means by which an organisation maximises the efficiency with which it plans, collects, organises, uses, controls, stores, disseminates, and disposes of its Information, and through which it ensures that the value of that information is identified and exploited to the maximum extent possible. The aim has often been described as getting the right information to the right person, in the right format and medium, at the right time

Information Technology
The technology used (e.g. applications and software) to support business functions and processes.

Knowledge
A combination of experience, values, information in context, and insight that form a basis for decisions making. It refers to the process of comprehending, comparing, judging, remembering, and reasoning;

Master data
Core reference data for the entities that an organisation owns or interacts with but not including transactional data

16 Glossary of terms

MDM
Master Data Management – The process of managing Master Data within an organisation which may be supported by purpose built software solutions

Measurement
Single activity to assess the state of a process, entity or system

Metadata
Generally defined as "data about data". Typically grouped into two types:
- Business metadata which explains meaning of data field and attributes;
- Technical metadata which will typically be used by database administrators to define data fields, for example number of characters, text/numeric format and validation rules

Monitoring
Ongoing measurement activities over time to maintain awareness of the state of a process, entity or system

Performance data
Assessment of the performance of an asset independently of its physical condition

Precision
The level that an attribute is recorded to, for example length can be measured in metres, millimetres or microns

Record
Evidence in the form of information representing an account of something that has occurred e.g. a maintenance record detailing an item of work being carried out;

Subjective information
Assessments of an asset, such as its condition and serviceability, based upon agreed assessment criteria;

Work management data
History of past activities, plans of future activities

17 Further reading

17.1 Bibliography

1. "Ensuring return on investment in Asset Information Systems", consultation workshop notes, Institution of Engineering and Technology, Warwick, November 2006
2. Extracted from Professor Thomas Davenport's paper "Putting the I in IT" from the Financial Times' Mastering Information Management Series, August 2002
3. BSI PAS 55-1:2008 Asset Management Part1: Specification for the optimized management of physical assets, BSI, September 2008
4. The International Infrastructure Manual (UK Edition) – available from Institute of Asset Management
5. Asset Information Guidelines 2001 (ISBN 0 7347 4104 9) – available on the Internet
6. Data Management for Road Administrations – A Best Practice Guide (Western European Road Directors - Sub-Group Road Data 2003)
7. "The Real Cost of Asset Information: How Better Costs Less", Ruth Wallsgrove, Sarras, 2003
8. Accenture US and UK Management Information Survey Findings, January 2007
9. Competency Requirements for the management of physical assets and infrastructure, The Institute of Asset Management, June 2006
10. "Practical Data Migration (Second Edition)", Johny Morris, BCS Books (ISBN 978-1-906124-84-7)

17.2 Publications

- BSI PAS55 Specification for the optimized management of physical assets, British Standards Institute
- IAM Competences Framework, Institute of Asset Management
- IAM Asset Information Guidelines, Institute of Asset Management
- Data Management for Roads Administrations, Western European Road Directors http://www.roaddata.org/data/sheets/RDBPG_Release_2.4.pdf
- "Six steps towards top data quality", Salesforce.com http://www.salesforce.com/community/crm-best-practices/administrators/data-management/data-quality/data-quality-best-practices.jsp
- Uniclass (Unified Classification for the Construction Industry), Construction Project Information Committee (CPIC)
- Omniclass - http://www.omniclass.org/
- Data Integrity Testing Best Practices, M.G.Harney
- "Achieving business benefits from improving the quality and integrity of location-based data", Digital National Framework http://www.dnf.org/documentation/library.asp?ID=DNF0052
- "Introduction to Data Integrity", Oracle http://download.oracle.com/docs/cd/B19306_01/server.102/b14220/data_int.htm#i3786
- "Improving information to support decision making: standards for better quality data", Audit Commission, 2007
- "How does Enterprise Data Management Relate to Data Warehousing?", Lance Miller, Teradata, Information Management & Data Quality Conference 2008
- "Enterprise Information Management and Data Governance for Business Leaders", Ladley. J, ImCue Solutions
- "Information Governance And Stewardship: Implementing Accountability for the Information Resource" Larry English, presentation, Brentwood, TN: Information Impact International, p. 17.
- "Improving Data Warehouse and Business Information Quality", Larry P. English, Wiley 1999
- "IS auditing guideline, IT governance, Document G18", ISACA (Information Systems Audit and Control Association"
- "IT Governance, Developing a successful governance strategy", National Computing Centre
- "TOGAF™ (The Open Group Architecture Framework)", The Open Group

17.3 Useful web sites

- Data Quality Pro – www.dataqualitypro.com
- Data Governance Institute - www.DataGovernance.com
- DAMA (Data Management International) - http://www.dama.org
- International Association for Information and Data Quality (IAIDQ) - http://iaidq.org/
- Method for an Integrated Knowledge Environment (MIKE2.0) –http://mike2.openmethodology.org
- OCDQ Blog - http://www.ocdqblog.com